Carbon-Related Materials

Camelia Miron • Paolo Mele
Satoru Kaneko • Tamio Endo
Editors

Carbon-Related Materials

In Honor of Nobel Laureate Akira Suzuki's Lecture at IUMRS-ICEM 2018

Springer

Editors
Camelia Miron
Center for Low-temperature Plasma Sciences
Nagoya University
Nagoya, Japan

Satoru Kaneko
Kanagawa Institute of Industrial Science
Kanagawa, Japan

Paolo Mele
Shibaura Institute of Technology
Tokyo, Japan

Tamio Endo
Japan Advanced Chemicals
Mie, Japan

ISBN 978-3-030-44229-3 ISBN 978-3-030-44230-9 (eBook)
https://doi.org/10.1007/978-3-030-44230-9

This Springer imprint is published by the registered company Springer Nature Switzerland AG
The registered company address is: Gewerbestrasse 11, 6330 Cham, Switzerland

Foreword

I am not a specialist in studies on carbon-related materials. I attended the International Conference of IUMRS-ICEM 2018 (International Union of Materials Research Societies—International Conference on Electronic Materials, Aug. 19–24, 2018, Daejeon, Korea), responding to the invitation by Prof. Tamio Endo (Japan Advanced Chemicals, Professor Emeritus Mie University) and Conference Committee (Prof. Soo Wohn Lee, Sun Moon University, Korea). I delivered Nobel Lecture on "Cross-Coupling Reactions of Organoboranes: An Easy Way for Carbon-Carbon Bonding." The special Symposium of "Carbon-Related Materials" was organized in commemoration of the Nobel Lecture. Some related photos are shown here. There I made many friends including Dr. Satoru Kaneko (KISTEC), Prof. Paolo Mele (Shibaura Inst. Tech.), and Prof. Soo Wohn Lee. This time, they requested me to write the preface for this book.

In 1963, I joined Professor Herbert C. Brown's research group, who received the Nobel Prize in Chemistry 1997, at Purdue University, IN, USA, as a postdoctoral associate, fascinated with the interesting new reaction, hydroboration. After 2 years of stay in Purdue University, I returned to Hokkaido University where I started to study for organic synthesis using organoboron compounds. We recognized the potential of organoboranes as intermediates in organic synthesis. Our discoveries of haloboration and cross-coupling reactions are fundamental contribution to the organic chemistry of boron and synthetic methodology.

The cross-coupling reaction is widely used for the stereodefined construction of carbon–carbon bonding in multifunctional systems. I retired from the university. However, I am very happy to have a chance to meet many young researchers at international meetings to discuss chemistry. I hope this book is useful for such young chemists.

Professor Akira Suzuki
HOKKAIDO UNIVERSITY PROFESSOR
Nobel Laureate in Chemistry, 2010
Hokkaido University, Sapporo, Japan

Scope of the Book

Functional carbon-based nanomaterials have become important due to their unique combinations of chemical and physical properties (i.e., thermal and electrical conductivity, high mechanical strength, and optical properties). Extensive research efforts are being made to utilize these materials for various industrial applications, such as high-strength materials and electronics. These advantageous properties of carbon-based nanomaterials are also actively investigated in several areas of microelectronic engineering.

This book offers the widest possible panorama of state-of-the-art in carbon materials and their applications, it is interdisciplinary and approach from engineering point of view. Are included detailed descriptions of different carbon-based materials synthesis methods, characterization, and industrial applications.

This book aims to be as much comprehensive as possible and will cover several categories of carbon materials such as graphene, carbon fiber composites, functionalized carbons, and polyimides. Different methods of synthesis are highlighted, such as CVD, plasma in liquids, fusion reactors, or frequency-doubled yttrium–aluminum–garnet (YAG) laser. Synthesis, characterization, and processing of superconducting materials is also covered.

Furthermore, the X-ray fluorescence (method that is integrated in the micro tomography and image processing laboratory) and Raman spectroscopy-based procedures are detailed.

Space will be dedicated to the important topics of microwave and slotted waveguide antennas based on carbon fibers as future applications, or studies concerning the response of azo-polymers to pulse laser irradiation, in the context of the surface relief gratings inscription. The state-of-the-art of polycondensation methods for various types of polyimide synthesis and their structural modification by plasma in liquids films will also be discussed.

This book will be invaluable to the experts to consolidate their knowledge and provide insight and inspiration to beginners wishing to learn about thermoelectric thin films. It is the ideal companion of the other two books on CRM we have edited.

Editorial Note

Dear Readers,

We are pleased to deliver the Special Book *Carbon-related materials—in honor of Nobel Laureate Akira Suzuki's lecture at IUMRS-ICEM 2018*, inspired from the Symposium "Carbon-related Materials: Carbon Nano Tubes, Diamond, Fibers, Graphene and Composites" during conference IUMRS-ICEM 2018 in Daejeon, Korea.

The authors of this book are distinguished colleagues and friends, some of them participants and contributors to symposium CRM in IUMRS-ICEM 2018. After almost two years, this book was finally edited and delivered today.

We would like to warmly thank our colleagues for their wonderful contributions, and everyone contributed with their precious help during the revision and editing of this book.

Camelia Miron, Nagoya University, Nagoya, Japan
Paolo Mele, Shibaura Institute of Technology, Tokyo, Japan
Satoru Kaneko, KISTEC, Ebina, Japan
Tamio Endo, Japan Advanced Chemicals, Mie, Japan

Introduction

This book originates from the Symposium "Carbon-Related Materials: Carbon Nano Tubes, Diamond, Fibers, Graphene and Composites" organized in the framework of the conference IUMRS-ICEM 2018 (International Union of Materials Research Societies—International Conference on Electronic Materials) from August 19th to 24th, 2018, at Daejeon Convention Center, in Daejeon, Korea http://www.iumrs-icem2018.org.

The symposium on carbon-related materials (in short CRM) was organized in honor of Nobel laureate Prof. Akira Suzuki, who delivered his Plenary talk "Cross-Coupling Reactions of Organoboranes: An Easy Way for Carbon-Carbon Bonding" on August 20, 2018.

The CRM symposium was organized by Helmut Takahiro Uchida, Tokai University, Japan; Masami Aono, National Defense Academy, Japan; Satoru Kaneko, KISTEC, Japan; Yasuharu Ohgoe, Tokyo Denki University, Japan; Yoku Inoue, Shizuoka University, Japan; and Tamio Endo, Japan Advanced Chemicals, Japan.

The CRM symposium provided multidisciplinary discussions for diverse carbon-related materials such as diamond, graphene, nanotubes, fullerene, amorphous carbon, and diamond-like carbon, including the formations, structures, properties, behaviors, and technological applications of CRM. Plastic and flexible materials were also included. The covered topics were: preparations, syntheses, and materials chemistry of CRM; Theoretical/computational approaches of CRM; Functionalization and surface modifications of CRM; Characterizations and properties of CRM; Relations between structures and properties of CRM; Device applications of CRM.

The contributions to CRM symposium were divided into three sessions "Carbon-based Devices Applications," "Carbon Nanotube Electronics," and "Deposition and Modification Methods for Carbon Materials" counting 11 talks (including 4 keynotes and 5 invited talks). High level of presentations and lively discussion contributed to the quality of the symposium.

Many outstanding oral and invited presentations were given during the CRM symposium. The symposium organizers were inspired by them to disclose such excellent papers to the widest scientific community. This is the reason why we invited our distinguished colleagues to share their results, and we publish this book entitled *Carbon-related materials—in honor of Nobel Laureate Akira Suzuki's lecture at IUMRS-ICEM 2018*.

We dedicate this book to Prof. Akira Suzuki in recognition to his outstanding career and constant inspiration to several generations of young chemists and scientists working on carbon-related materials.

This book is intended as third book of the series *Carbon-related materials in honor of Prof. Suzuki* after the first (related to the conference ICCE-23 and ICCE-24) published by Springer in 2017 and the second (related to the conference IUMRS-ICAM 2017) published by Springer in 2020. The three books are all edited by the same core team of researchers with the endorsement of THM (Team Harmonized Materials) Japan.

April 30th, 2020

Camelia Miron—Nagoya, Japan
Paolo Mele—Tokyo, Japan
Satoru Kaneko—Ebina, Japan (organizer of the CRM symposium)
Tamio Endo—Mie, Japan (organizer of the CRM symposium)

Contents

Contributors

Octavian Baltag "Grigore T. Popa" University of Medicine and Pharmacy of Iasi, Iasi, Romania

William Baron Air Force Research Laboratory, WPAFB, OH, USA

Alexe Bojovschi Center of Excellence for Technology Innovation, IntAIB, Melbourne, VIC, Australia

Department of Electronics and Communication Engineering, ASET, Amity University, Noida, India

Robert A. Canfield Department of Aerospace and Ocean Engineering, Virginia Tech, Blacksburg, VA, USA

Takahiro Ishizaki Department of Materials Science and Engineering, Shibaura Institute of Technology, Tokyo, Japan

Martin Kalbac Department of Low-Dimensional Systems, J. Heyrovsky Institute of Physical Chemistry, Czech Academy of Sciences, Prague, Czech Republic

Woon Kim Department of Aerospace and Ocean Engineering, Virginia Tech, Blacksburg, VA, USA

Geoffrey Knott Department of Mechanical Engineering Sciences and Surrey Space Centre, University of Surrey, Guildford, UK

Juergen F. Kolb Leibniz-Institute for Plasma Science and Technology, INP Greifswald, Greifswald, Germany

Tu C. Le School of Engineering, RMIT University, Melbourne, VIC, Australia

Cristian P. Lungu National Institute for Laser, Plasma and Radiation Physics, Bucharest, Romania

Jason Miller Booz Allen Hamilton, Dayton, OH, USA

Camelia Miron Center for Low-temperature Plasma Sciences, Nagoya University, Nagoya, Japan

Kelvin J. Nicholson Aerospace Composite Technologies—Aerospace Division, Defence Science and Technology Group, Melbourne, VIC, Australia

Georgiana Rosu "Ferdinand I" Military Technical Academy, Bucharest, Romania

Liviu Sacarescu "Petru Poni" Institute of Macromolecular Chemistry, Iasi, Romania

Ion Sava "Petru Poni" Institute of Macromolecular Chemistry, Iasi, Romania

Iuliana Stoica "Petru Poni" Institute of Macromolecular Chemistry, Iasi, Romania

James Tuss Air Force Research Laboratory, WPAFB, OH, USA

Jana Vejpravova Department of Condensed Matter Physics, Faculty of Mathematics and Physics, Charles University, Prague, Czech Republic

Andrew Viquerat Department of Mechanical Engineering Sciences and Surrey Space Centre, University of Surrey, Guildford, UK

Chapter 1
Introduction to Raman Spectroscopy of Chemically Functionalized CVD Graphene

Jana Vejpravova and Martin Kalbac

1.1 Introduction

Single-layer graphene (1-LG) is the archetype two-dimensional (2-D) material. It comprises a one-atom-thick layer of carbon atoms, which are assembled into a honeycomb-like lattice [1]. Because of its fascinating properties [2–4], the graphene has inspired many experimental and theoretical works underlying a plethora of popular applications of graphene [5–9]. As the 1-LG represents a standalone surface, its interaction with surroundings is essential factor influencing its band structure (position of the Fermi level and distortion of the Fermi surface, etc.) thus its physiochemical properties in general.

High quality samples of graphene can be obtained by mechanical exfoliation of the flakes from graphene. However, from a practical application's point of view, it is more important to obtain a large area of the material [10]. Graphene can be prepared by chemical exfoliation from bulk materials [11]. The procedure is typically quite harsh and therefore one obtains material that is purely defined and has a significant amount of defects. This procedure also leads to the presence of multilayers, and the distribution of flake size could be quite broad [12]. Therefore, such a material is not suitable for high-end applications.

The chemical vapor deposition (CVD) is a compromise between mechanical and chemical exfoliation approaches [13, 14]. In the case of graphene, the CVD process

J. Vejpravova (✉)
Department of Condensed Matter Physics, Faculty of Mathematics and Physics,
Charles University, Prague, Czech Republic
e-mail: jana@mag.mff.cuni.cz

M. Kalbac
Department of Low-Dimensional Systems, J. Heyrovsky Institute of Physical Chemistry,
Czech Academy of Sciences, Prague, Czech Republic
e-mail: martin.kalbac@jh-inst.cas.cz

C. Miron et al. (eds.), *Carbon-Related Materials*,
https://doi.org/10.1007/978-3-030-44230-9_1

can lead to the samples of relatively high quality. Taking advantage of the fact the CVD for graphene is already in the advanced stage, one can consider using the graphene as a starting material for further modification, which will alter its properties. Although the research on pristine graphene is not exhausted yet, there is also growing interest in chemical functionalization of graphene [15]. In that sense, one can think of various modifications broadening the number of potential applications depending on specific functional groups attached to the graphene.

In general, chemical functionalization of graphene covers a quite broad range of approaches giving rise to either non-covalent or covalent interaction between the graphene and a coveted functional group or a molecule [15–19]. In fact, the close-to-ideal graphene is quite inert; therefore, harsh conditions and strong reagents are often needed, which usually causes serious damage to the graphene sample [20].

Considering that only a sub-monolayer of functional groups is anchored to a modified graphene surface, it is very challenging to prove their presence and to determine the nature of the chemical bond between the chemical species and the graphene [21]. The most convenient technique for studying graphene is Raman spectroscopy (RS). In the case of chemically modified mono- to few-layer graphene samples, direct detection of the functional groups is impossible; however, some indirect fingerprints can be derived from the RS. In this chapter, the current state of the art for studying chemical functionalization of CVD graphene by RS will be presented.

The chapter is structured as follows. After a brief Introduction (Sect. 1.1), a summary of RS of graphene is given in Sect. 1.2, including the effects of doping, strain, stacking, and defects on the Raman spectra (RSp). Section 1.3 summarizes procedures for the chemical functionalization of CVD graphene and further focuses on the methodology of detection using RS. In particular, the use of surface-enhanced RS (SERS) for studying the CVD graphene and functionalized CVD graphene is discussed. Conclusions and outlooks are given in Sect. 1.4.

1.2 RSp of CVD Graphene

RS is the most widely used tool to study and characterize graphene samples [22, 23]. It allows distinguishing between one- and two-layered graphene, it can provide information about defects in graphene, about doping of graphene or mechanical strain in graphene. Many works report on details of the RS used for graphene investigations, i.e., [22–25]. Therefore, only a summary targeting readers not familiar with the RS on graphene is given in this section. Besides, the use of carbon isotope labeling as a very important tool in the identification of chemical functionalization on double- and few-layer CVD graphene is introduced.

1.2.1 Phonon Processes in Graphene

Figure 1.1a shows a typical RSp of graphene, which consists of four important bands: the G and the 2D mode (also called G′ mode in the literature), and defect-related D and D′ modes. The G band in the RSp of pristine graphene is found at about 1580 cm^{-1}. The D and 2D (G′) modes are observed in the spectral regions of 1250–1450 cm^{-1} and 2500–2900 cm^{-1}, respectively.

The origin of the bands will be now briefly discussed. The unit cell of graphene contains two carbon atoms, which gives six phonon modes. Of those modes, three are acoustic (A) and three optical (O). Both acoustic and optical phonon branches consist of one out-of-plane vibration (o) and two in-plane vibrations (i). The in-plane vibration can be parallel (L) or perpendicular (T) to the line connecting two nearest carbon atoms.

Because of the phonon momentum conservation requirement, the first-order Raman features originate from the close vicinity of the Γ point in the first Brillouin zone of graphene. The iTO and iLO phonon branches merge at the Γ point and give rise to the G mode of graphene. In other words, the G mode originates from a double generate (E_{2g}) phonon mode.

The physical origin of 2D and D modes has been explained by double-resonance theory [26]. The selection rules for electron–phonon scattering allows activation of the iTO phonon connecting electronic states near the K and K′ points of the Brillouin zone. This intervalley process gives rise to the 2D mode.

The one-phonon second-order Raman D band appears if there is a breakdown in translational crystal symmetry, which can be caused by defects in the structure. On the other hand, the two-phonon second-order Raman 2D feature occurs independently of the presence of structural defects. The D mode is important for the quantification of defects in graphene [24, 27].

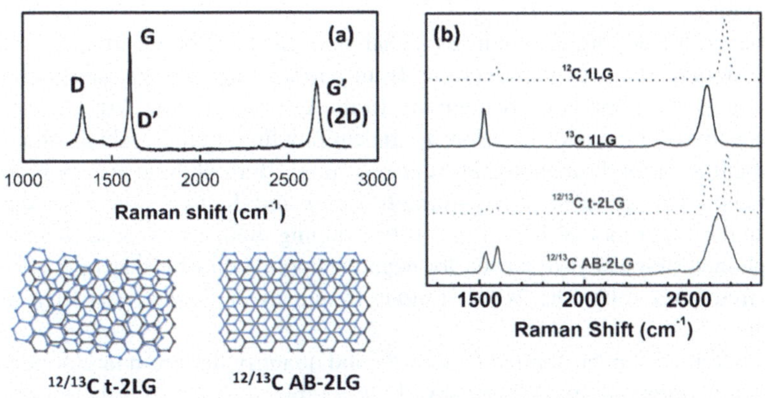

Fig. 1.1 RSp of ^{12}C CVD graphene with the assignment of the important Raman active bands (**a**), and comparison of ^{12}C, ^{13}C 1-LG and isotopically labeled T-2-LG and AB-2-LG (**b**). The two types of stacking (turbostratic and AB) are also shown

Another defect-related mode—D′ originates from the intravalley process. The D′ mode is located at about 1610 cm^{-1}; hence, it is overlapping with the G mode, especially in the case that the sample is doped. Therefore, the D′ mode is typically not analyzed unless the number of defects is large.

Because of the double-resonance nature of D and 2D modes, their Raman signal reflects both the electronic structure of graphene and phonon dispersion relations in graphene, consequently both the D and the 2D Raman mode exhibit dispersive behavior. Hence, when a different laser excitation energy is used, a phonon with a different q vector and different energy is employed in the resonance process.

1.2.2 Doping and Strain Effects on RSp of CVD Graphene

CVD graphene samples are usually studied on substrates. The substrates can cause strain and doping of graphene, and in some cases, there is even a possible mixing of electronic states of graphene and the specific substrates. Therefore, the substrate plays a crucial role in the studies of graphene and 2-D materials in general.

Both the G and the 2D modes frequencies shift and change their intensity upon doping. The frequency shift of the G band in charged graphene is related to the change in the C–C bond strength and to the renormalization of the phonon energy [28]. In graphene, a coupling between lattice vibrations and Dirac fermions is allowed because the scales for the electron and phonon dynamics are comparable. Therefore, the adiabatic Born–Oppenheimer approximation fails to describe the G band phonons and time-dependent perturbation theory is needed to explain the experimental observations. In charged graphene, the Fermi energy E_F is moved away from the Dirac point, and thus the formation of electron–hole pairs is suppressed [28]. Because of electron–hole symmetry with respect to the Dirac point, the frequency shift of the G mode should be identical both for positive and negative doping.

However, the doping also induces a change of the C–C bond strength [29]. The positive doping removes the electrons from antibonding orbitals, and therefore a hardening of the phonon corresponding to the G band is expected. On the other hand, negative doping adds electrons to the antibonding orbitals, which should lead to a softening of the Raman signal frequency (ω_G). Both phonon energy renormalization and a change of the bond strength occurs and the two effects are superimposed in the experimental RSp. For positive doping, both effects lead to an upshift of the phonon frequency. However, for negative doping, they have an opposite effect on the frequency shift. The 2D (G′) mode frequency ω_{2D} is also sensitive to doping [30].

In general, RS can be applied to identify and quantify the strain in graphene. The strain can be uniaxial, biaxial or triaxial, or combined [3, 31]. Furthermore, it has been demonstrated that strained graphene can mimic the electronic structure like it would be placed in a giant magnetic field [4]. The strain manifests itself both in the frequency of the G and 2D modes. Larger changes are observed in the case of the 2D mode. Nevertheless, also the 2D mode is dependent on the doping; therefore, one needs to decouple the effects of the doping and the strain.

As discussed above, the G mode frequency is very sensitive to doping and less sensitive to strain; on the other hand, the 2D mode frequency is very sensitive to strain and only slightly sensitive to doping. Thus, it is possible to extract the changes in strain and doping in graphene samples even if both effects are present simultaneously. For a proper analysis of strain and doping in graphene samples, it is necessary to measure Raman maps and from the obtained data construct correlation plots of the 2D and G mode frequencies [32, 33].

1.2.3 RSp of Graphene Multilayers

The multilayer graphene represents a step toward more complex 2-D materials. The simplest and most widely studied representative of this new class of 2-D materials is bilayer graphene. As compared with single-layer graphene, the bilayer graphene has a new free parameter, which is the orientation of the layers with respect to each other. In general, random orientation is called turbostratic bilayer (T-2-LG), and the special case when the carbon of one layer are located just above the center of the hexagons of the second layer is called AB-stacked or Bernal-stacked bilayer (AB-2-LG). The orientations of the layers reflect in the changes in the electronic structure. This change is most obvious for the AB-stacked 2-LG.

Nevertheless, as was demonstrated recently, also the specific orientation of the layers leads to the formation of van Hove singularities in the electronic structure of graphene [34, 35] or even to superconducting or Mott insulating states [2]. Because the orientation of graphene layers can be arbitrarily set, these results open a path to realizing devices with tunable electronic structure.

The multilayers of graphene with a defined number of layers can be prepared by subsequent transfer of one graphene layer on top of another graphene layer(s) [36]. The bilayers prepared by the transfer method are typically turbostratic as they do not follow the order of layers in graphite. Recently, it was shown that the angle between the layers can be finely tuned by using a specialized transfer stage. For AB-stacked graphene, one can also apply the modified condition of the CVD protocol, which leads to the formation of bilayer/multilayers.

RS can be used to distinguish between 1-LG and AB-2-LG; however, the only difference is broadening and change in symmetry of the 2D mode of the AB-2-LG as compared with the 1-LG. In the case of T-2-LG, the spectra typically do not differ from the spectrum of 1-LG. However, for some specific angles between graphene layers, one can observe a strong enhancement of the Raman signal due to resonance with van Hove singularities [34]. Using the so-called isotope labeling by ^{13}C [25], the Raman frequency of graphene features are shifted to lower frequencies following Eq. (1.1):

$$(\omega_0 - \omega)/\omega_0 = 1 - \left[(12 + c_0)/(12 + c)\right]^{1/2} \tag{1.1}$$

where ω_0 is the frequency of a particular Raman mode in the ^{12}C sample, $c = 0.99$ is the concentration of ^{13}C in the enriched sample, and $c_0 = 0.0107$ is the natural abundance of ^{13}C.

Now, if we compose the 2-LG from the ^{12}C layer and ^{13}C layer, it is possible to distinguish the top and the bottom layers of 2-LG. Isotope labeling thus enables a deeper understanding of the chemical functionalization of graphene bilayers, which will not be possible without this technique.

The stacking order of graphene layers in the graphene bilayer can also be confirmed by RS when isotope labeling is applied. Figure 1.1b shows RSp of turbostratic 2-LG and the AB-stacked 2-LG. For both AB and turbostratic graphene layers, one can find two G modes, corresponding to the top and bottom graphene layers. However, the appearance of the G′ (2D) mode is different. While in the case of turbostratic 2-LG there are two modes, in the case of AB-stacked 2-LG only one broad band is found. This can be rationalized by considering the number of phonons involved in the Raman process [37].

1.2.4 *Features of Defects in RSp of CVD Graphene*

As in conventional semiconductors, defects in graphene strongly influence its properties and may have either detrimental or overall beneficial effects on the characteristics of the material. Examples of the former are a decrease in electron mobility or a drop in mechanical characteristics with an increase in defect concentration [38, 39]. In general, many types of defects can be found in graphene including structural (sp^2-like) defects, topological (sp^2-like) defects, doping or functionalization (sp^2- and sp^3-like) defects, and vacancies/edge type defects (non-sp^2-like).

The defects have also characteristic fingerprints in RSp of graphene. The type and number of defects can be evaluated, i.e., by monitoring the ratio of D/G and D/G′ mode intensities [24, 27, 40, 41].

The evaluation of defects is also very important for the chemical functionalization of CVD graphene as any covalent or non-covalent interaction may be viewed as a kind of defect. Consequently, the most common way of evaluating the level of chemical functionalization is via monitoring the intensities of the D, D′ comparing to the G and 2D.

Another important problem is related to the size of the area affected by the defects, which plays an important role mainly for multilayer graphene samples.

In the case of 1-LG, it was shown that the impact of defects can be seen up to about 1.8 nm away from their location [41]. As the distance between graphene layers in a bilayer is about 0.335 nm only one can raise a question, whether the defect is also affecting nearby layers. However, the evaluation of the location of defects is in general quite difficult because the position of the D mode is the same for both graphene layers. However, with the help of isotope labeling, one can distinguish between the defects in the top and the bottom layers [25, 37].

1.3 Raman Spectroscopy of Functionalized CVD Graphene

1.3.1 Chemical Functionalization of CVD Graphene

Graphene, in general, can be functionalized by many different routes, but most of them are realized in the liquid phase [18]. As the graphene surface can be easily contaminated even if ultra-high purity solvents are used most of the approaches developed primarily for chemically exfoliated graphene-like flakes [15, 18] are not suitable for functionalization of CVD graphene. Even though the majority of works dealing with functionalization of CVD graphene rely on diazonium chemistry [20, 42], which is somewhat risky through the defect formation or even layer removal or full damage. The CVD graphene also responds differently to most of the chemical processes working in the liquid phase as its reactivity is strongly affected by the substrate [42–44], topographic corrugations, and doping [45]. So far click chemistry protocols [17] as well as various reactions in the gas phase, like hydrogenation [46, 47] or fluorination [48–50] were successfully carried out on CVD graphene. Among them, the most convenient reaction seems to be fluorination using XeF_2 [48]. Moreover, the fluorinated CVD graphene is a perfect starting point for a mild introduction of various organic groups [51, 52], as shown in Fig. 1.2.

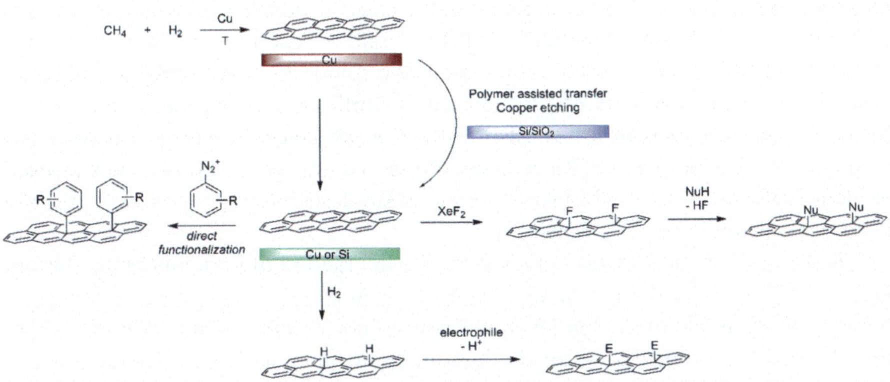

Fig. 1.2 General approach of anchoring various chemical species onto CVD graphene

1.3.2 Raman Spectroscopy of Fluorinated CVD Graphene

Fluorinated graphene termed often fluorographene was first prepared by Nair and coworkers [48] using XeF_2 at slightly elevated temperatures. The resultant product showed thermal stability limited to 400 °C in an atmospheric environment. RS study revealed enhancement of the defect-related features dominated by the continuous increase in the D band intensity at 1350 cm^{-1} and suppression of the 2D band indicating an increase of the fluorination degree. Fluorination was attained until all D, 2D, and G bands disappeared.

Furthermore, fluorinating CVD graphene on both the front and back surface was achieved on Si substrate only because of the effective etching on the Si substrate by XeF_2 gas [53]. This effect was confirmed by fluorinating graphene on Cu foils because the Cu substrate cannot be extensively etched by XeF_2. RSp also indicated the introduction of a high degree of structural disorder in the graphene after fluorination [53].

Fluorination using XeF_2 has also been carried out on 2-LG, which can serve as a model system to study the role of the substrate in graphene reactivity as the 2-LG can be viewed as a 1-LG on a graphene substrate [37]. Furthermore, one can also evaluate the role of the specific orientation of graphene with respect to the substrate by comparing the reactivity of T-2-LG and AB-2-LG. To avoid the interference of different effects and contamination, one needs to select an appropriate test reaction, which is the "clean fluorination" using XeF_2. To ensure the same conditions for comparison of the reactivity of graphene on different substrates, a series of samples was fabricated (single-layer graphene, T-2-LG, and AB-2-LG all placed on same Si/SiO_2 substrate).

As already discussed, the reactivity of graphene can be easily evaluated by RS because the higher fluorination level corresponds to the stronger D mode in the RSp. The comparison of the RSp before and after fluorination is shown in Fig. 1.3. The highest reactivity exhibits 1-LG, which is demonstrated by the strong D mode after the reaction. Smaller reactivity is found in the case of the T-2-LG, which represents a "graphene on graphene substrate" [54]. The smallest reactivity is found in the case of the AB-2-LG. These results demonstrate the importance of graphene–substrate orientation for the reactivity of graphene. It is worth mentioning that in the case of 2-LG, the D mode was observed also in the bottom layer. However, this does not mean that the bottom layer is fluorinated. (Note that graphene fluorination was used to identify the position of add-layers grown on single-layer graphene as only the top layer is expected to be fluorinated [36].)

The D mode in these cases comes from the interaction of phonons in the bottom layer with defects in the top layer. Thanks to isotope labeling, we can distinguish between the D mode of the top layer and bottom layer, consequently, the results are not affected by a contribution of the bottom layer phonon–defects interactions. In the case of the 2-LG, the isotope labeling also enables disentangling the effects of the functionalization on doping of the top and bottom layers. These effects, however,

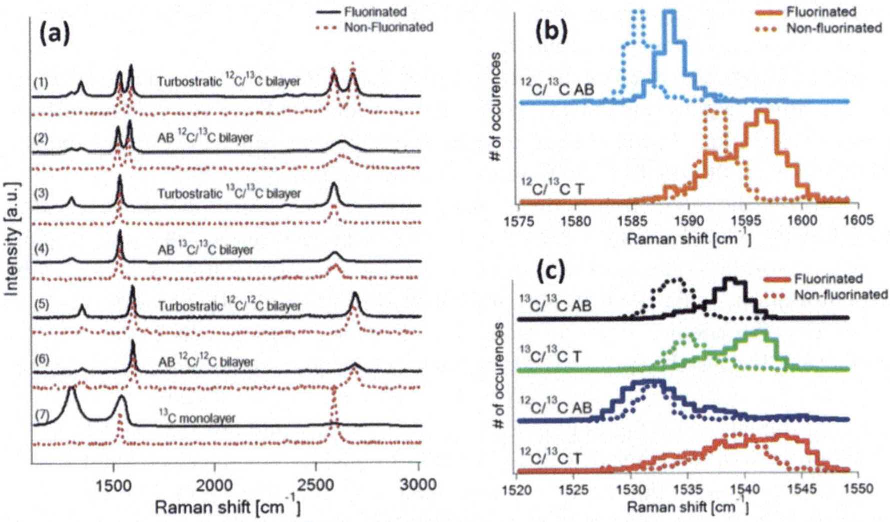

Fig. 1.3 RSp of graphene subjected to fluorination. (**a**) RSp of the 1-LG, and isotopically labeled of the T-2-LG and AB-2-LG before and after fluorination. (**b**) Histograms of the G mode frequencies of the ^{12}C top layer for the non-fluorinated and fluorinated isotopically labeled 2-LG. (**c**) Histograms of the G mode frequencies of the ^{13}C bottom layer for the non-fluorinated and fluorinated isotopically labeled 2-LG. The figure was adopted from [54]

have to be studied using Raman mapping to avoid local Inhomogeneities and to enable rigorous correlation of the spectral parameters.

Figure 1.3b shows the distribution of the G mode frequencies of the top graphene layer before and after fluorination. The frequency of the G mode is clearly increased after fluorination. This is in line with the expectation as the fluorine functional group is withdrawing electrons from graphene. The effect is slightly more pronounced for the T-2-LG, which corresponds to higher fluorine content. Figure 1.3c shows the distribution of the G mode frequencies for the bottom graphene layer. In the case of the T-2-LG, the G mode of the bottom layer is upshifted similarly to that of the top layer. On the other hand, for the AB-2-LG, there is almost no shift. This resembles the behavior of the AB-2-LG during the electrochemical doping and points to almost no charge transfer between the top and bottom layers in AB-2-LG [37].

1.3.3 SERS on CVD Graphene

A Si substrate with a metal oxide layer of a specific thickness is usually employed to enhance the Raman signal, which is called interference-enhanced RS (IERS) [55]. In particular, its enhancement of nearly 30 times for a graphene monolayer on a Si wafer with a 300 nm thick SiO_2 layer makes this substrate the most common in graphene research. However, when the graphene is placed on an arbitrary substrate, the Raman signal decreases dramatically.

A facile approach to enhancing the signal is the SERS [56]. When a few nanometer thin film of a plasmonic metal (gold, silver) is deposited on the graphene surface, up to 10^3 signal enhancement can be achieved depending on the excitation energy, film thickness, morphology, and plasmonic material [48, 57–60].

However, the interaction of the graphene with the plasmonic metal causes additional changes to the RSp, which are usually dispersive due to an intrinsic plasmonic absorption characteristic of a particular plasmonic film. Consequently, decomposition of the RSp is not straightforward [57, 58, 60, 61], i.e., one can observe the broadening of the G mode thanks to the complex light–matter interaction [60]. It has also been observed that plasmonic nanoantennas tuned to the Stokes wavelengths associated with the G and 2D Raman bands of graphene not only enhance Raman scattering in graphene but also displace and broaden the RSp [62].

Figure 1.4 shows typical RSp for a CVD graphene monolayer covered with a 10 nm silver or gold thin films in comparison to the pristine 1-LG. The RSp were measured at different excitation wavelengths. It is clear that the deposition of metal on graphene leads to substantial deviations in the RSp. Explicitly, there is a change in the Raman shifts, FWHM, and intensity for both the G and the 2D mode. The G mode broadens and shifts toward lower energies. It is also obvious that the broadening is much larger for the gold layer within the interval of the Raman excitation wavelengths, and it significantly depends on the laser excitation energy for both the gold and silver layers. The maximum broadening for the silver seems to be at the

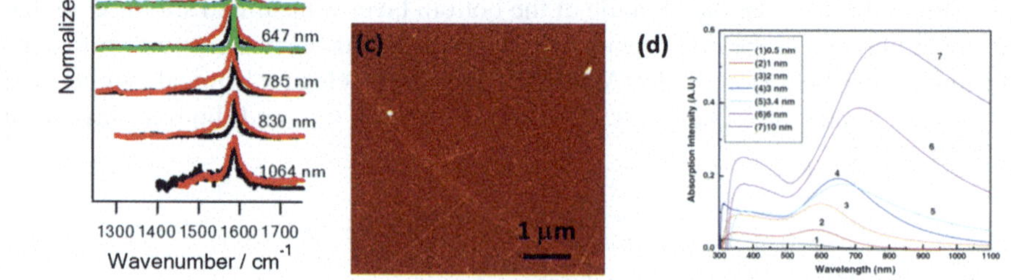

Fig. 1.4 Evolution of the G mode in RSp of the graphene covered with the gold (red) and silver (black) 15 nm layer together with the bare (green) graphene (**a**). All spectra are normalized to the maximum intensity in the corresponding region. Panel (**b**) shows the distance-dependent SERS enhancement for isotopically labeled graphene bilayers. Typical topography of the plasmonic films on graphene obtained by AFM is shown in panel (**c**). Evolution of the plasmonic absorption characteristics for the different thickness of Au films is given in panel (**d**). The figure was reworked using results published in [61, 64]

514 nm excitation line, whereas for the gold layer, it is between 647 and 785 nm. This coincides very well with the plasmonic characteristics of the silver and gold thin films [63]. Apart from the signal enhancement, a significant change in the position and FWHM of the Raman bands is present, and the fine structure of the G mode can be resolved. The influence on the RSp is stronger when the laser excitation energy coincides with the plasmon frequency of the metal. In the case of gold, stronger interactions with graphene are expected in contrast to the silver metal. The observed phenomena can be explained by the synergy of two effects. First, the silver-coated sample quickly creates an oxidized/sulfonized surface layer, thus causes decoupling the metal and the graphene. Second, the surface plasmon polariton–graphene interaction dominates for the gold-covered sample in the interval of the excitation energies used in the experiments. The slope of the 2D mode dispersion is also affected by the metal–graphene interaction, and it is larger for the gold-covered sample. In contrary to the much weaker interaction of the graphene with the silver layer, this system is much more convenient for application of the SERS for studies of functionalized single-layer graphene as the signal enhancement is much larger, while the character of the graphene bands is less influenced [61].

The SERS can also be applied to bilayer and few-layer graphene [58]. However, the experimental enhancement factor of SERS in the 1-LG is higher in the G band than in the 2D band, whereas in the few-layer graphene it is nearly identical in the G band and the 2D band. This indicates that the G band in the single-layer graphene is susceptible to SERS. Such a trend in the enhancement factors is also reflected in the ratio of the Raman intensities of 2D and G band (I_{2D}/I_G), which for single-layer decreases by ~50% after deposition of plasmonic films. The I_{2D}/I_G value in the few-layer graphene is less affected [58].

Recently, isotopically labeled graphene bilayers were used to study the effects of the distance on SERS enhancement, which is not possible to address in common few-layer samples. Figure 1.3b shows the sandwich heterostructure consisting of the ^{12}C graphene layer, ^{13}C graphene layer, and 15 nm of the gold layer (AFM topography presented in Fig. 1.3c), together with the RSp of these sandwiches. The overall signal is significantly enhanced. However, as can be observed in Fig. 1.3b, the signal of the graphene layer closer to the gold layer is stronger than the signal of the layer, which is further away. This is in contrast to the uncovered graphene, where the intensity of the G mode of the top and bottom layers is almost identical.

A more detailed analysis shows that the change of the signal intensity matches the decrease of the electromagnetic field generated by a plasmon at the distance of about 0.34 nm from the layer next to the plasmonic layer according to the distance dependence of the intensity of the SERS signal, which can be calculated within the E^4 approximation according to Eq. (1.2):

$$I(r) = \left(\frac{a+r}{a}\right)^{-10}, \qquad (1.2)$$

where a is the radius of curvature of the field-enhancing features and r is the surface–adsorbate distance [56].

It has also been demonstrated that morphology of the plasmonic film (particle size and shape) are important factors in SERS on graphene [65–67].

1.3.4 Surface-Enhanced RSp of Functionalized CVD Graphene

Being aware of some difficulties in interpretation of SERS spectra of graphene discussed in the previous section, the approach has been identified as very powerful for direct observation of various chemical species anchored to CVD graphene. The development of the functionalization protocols on substrates different from Si/SiO_2 is complicated because characterization methods for graphene or functionalized graphene are limited in number, sensitivity, and experimental conditions. As discussed above, RS is routinely used to investigate graphene functionalization due to the emergence of the D mode attributed to the creation of sp^3 "defects" in the sp^2 carbon lattice. However, this feature is not indicative of the actual chemical nature of the functionalization. Among others, SERS is particularly promising as it may reveal the characteristic Raman fingerprints of the grafted functional groups.

The first attempt to use SERS on chemically functionalized graphene was applied to graphene on copper [52]. The fluorination was performed using XeF_2 in the gas phase and the subsequent nucleophilic exchange by S-, N-, and O-nucleophiles was performed. As a benchmark, we have chosen a reaction of fluorinated graphene with thiophenol in the gas phase to undergo nucleophilic substitution of fluorine atoms by sulfur (phenylsulfanyl group) [51]. Pristine graphene on Cu measured with 633 nm excitation laser shows the characteristic G and 2D bands at 1586 and 2658 cm^{-1}, respectively. Fluorination leads to the appearance of the D mode at 1343 cm^{-1}, which is accompanied by a dramatic decrease in the 2D mode intensity and shift of the G band to 1596 cm^{-1}. After the reaction with thiophenol in the gas phase, the 2D/D intensity ratio increases, and the D' band (1621 cm^{-1}) is resolved from the G mode, which shifts to lower wavenumber (1584 cm^{-1}). In the case of functionalization on copper, the characteristic phenylsulfanyl bands have been observed in the spectrum due to the plasmonic enhancement by the substrate, while on Si/SiO_2 these bands appeared only after deposition of silver film and measuring using SERS [51, 52]. In general, the RSp measured on graphene on copper show the typical graphene modes as well as characteristic vibrations of the functional groups if Raman active.

Figure 1.5 shows an example of covalent functionalization of CVD graphene on SiO_2/Si substrate with thiophenol [51] (the reaction scheme is given in Fig. 1.5d). A typical RS of the starting material is shown in Fig. 1.5a; the dominant G and 2D modes and lack of D and D′ suggest a negligible amount in defects. After fluorina-

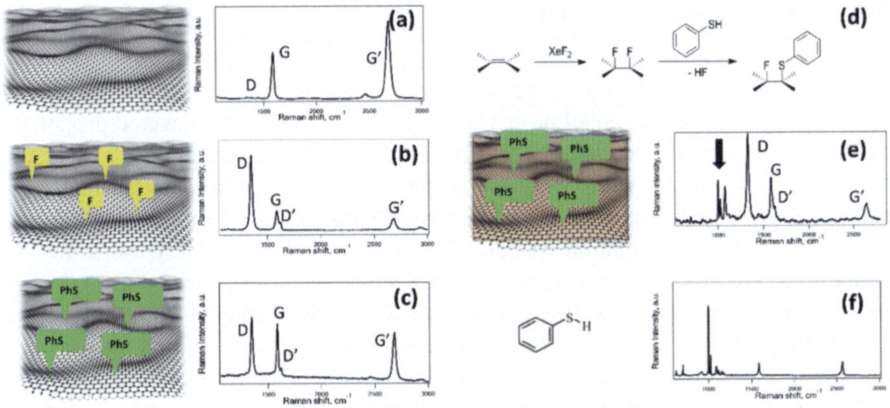

Fig. 1.5 Detection of chemical functionalization of CVD graphene at each reaction step using RS and SERS. Pristine graphene (**a**) first undergoes fluorination (**b**) and exchange of fluorine to phenylsulfanyl (**c**) according to the reaction scheme presented in panel (**d**). Direct observation of the functionalization with the help of SERS is shown in panel, the black arrow marks the position of the phenylsulfanyl fingerprints (**e**). RS of pure thiophenol is shown in panel (**f**)

tion using XeF_2, significant D and D′ modes appear suggesting that the fluorination process was successful (Fig. 1.5b). However, one cannot conclude unambiguously that the defect-related modes are present due to C–F bond or just due to defects created during the fluorination. After exchanging the fluorine to phenylsulfanyl (Fig. 1.5c), the RSp does not change dramatically. At this point, one can hardly decide whether the exchange process underwent with partial or full yield, or did not take place at all. At this stage, additional techniques like X-ray photoelectron spectroscopy are usually applied and even though there is no unambiguous evidence of covalent coupling of the molecule.

This information can be obtained when one profits from the SERS effect. After covering the functionalized CVD graphene by a plasmonic film, the RSp changes dramatically as the features intrinsic to the phenylsulfanyl appear (Fig. 1.5e). When compared to the spectra of pure thiophenol (Fig. 1.5f), a Raman active mode of the ν S-H stretching at around 2600 cm^{-1} is no more present in the RSp of functionalized graphene (Fig. 1.5e) suggesting covalent coupling of the phenylsulfanyl to graphene.

One can easily expand the SERS detection to functionalized bilayer and few-layer graphene and simultaneously profit from the isotope labeling of the individual layers. It is essential for further investigations of the nature of coupling various species with unique RSp features to the CVD graphene monolayers, bilayers, and few layers.

1.4 Conclusions and Outlook

Graphene and 2D materials in general, are base components for the building of new classes of materials with novel properties. The ability to tune the coveted properties of these components by chemical functionalization and addressing precisely the resulting electronic structure of these components is expected to contribute to a revolution in nanoscience and nanotechnology in upcoming years.

Advanced RS investigation was demonstrated to be an extremely useful tool to study the CVD graphene. The capability of RS can be further boosted to studies of multilayer systems by using the isotope labeling, which enables addressing individual graphene layers. The SERS on chemically functionalized graphene has been identified as a powerful tool for direct observation of various chemical species anchored to the CVD graphene monolayers and bilayers, thus gives information on the character of the chemical bond, which is almost impossible to address by any other technique. The universality of the approach is given by its outreach to many different chemical reactions, which can be further expanded to other 2D materials. In the latter case, the 2D material(s) may not respond strongly to the external effects (like doping), but the graphene can be eventually placed on top or beneath them can serve as a probe, which conveniently senses these variations. Further advancements in chemical functionalization of graphene, i.e., reaching a submicrometer spatial resolution, inter-linking with other 2D materials including a reversible coupling will promote considerable progress in the fundamental and technological playground of 2D material's applications.

Acknowledgements This work was supported by European Research Council (ERC-Stg-716265), Czech Science Foundation (18-20357S) and Ministry of Education, Youth and Sports of the Czech Republic under Operational Programme Research, Development and Education (project CARAT, CZ.02.1.01/0.0/0.0/16_026/0008382). We acknowledge P. Kovaricek from J. Heyrovsky institute of Physical Chemistry, Prague for providing general scheme of chemical functionalization.

References

1. Novoselov KS, Geim AK, Morozov SV et al (2005) Two-dimensional gas of massless Dirac fermions in graphene. Nature 438:197–200. https://doi.org/10.1038/nature04233
2. Cao Y, Fatemi V, Fang S et al (2018) Unconventional superconductivity in magic-angle graphene superlattices. Nature 556:43
3. Si C, Sun Z, Liu F (2016) Strain engineering of graphene: a review. Nanoscale 8:3207–3217. https://doi.org/10.1039/C5NR07755A
4. Levy N, Burke SA, Meaker KL et al (2010) Strain-induced pseudo–magnetic fields greater than 300 tesla in graphene nanobubbles. Science 329:544–547. https://doi.org/10.1126/science.1191700
5. Yang K, Wang J, Chen X et al (2018) Application of graphene-based materials in water purification: from the nanoscale to specific devices. Environ Sci Nano 5:1264–1297. https://doi.org/10.1039/C8EN00194D

6. Fiori G, Bonaccorso F, Iannaccone G et al (2014) Electronics based on two-dimensional materials. Nat Nanotechnol 9:768
7. Koppens FHL, Mueller T, Avouris P et al (2014) Photodetectors based on graphene, other two-dimensional materials and hybrid systems. Nat Nanotechnol 9:780
8. Kostarelos K, Novoselov KS (2014) Graphene devices for life. Nat Nanotechnol 9:744
9. Han W, Kawakami RK, Gmitra M, Fabian J (2014) Graphene spintronics. Nat Nanotechnol 9:794
10. Yi M, Shen Z (2015) A review on mechanical exfoliation for the scalable production of graphene. J Mater Chem A 3:11700–11715. https://doi.org/10.1039/c5ta00252d
11. Xu Y, Cao H, Xue Y et al (2018) Liquid-phase exfoliation of graphene: an overview on exfoliation media, techniques, and challenges. Nano 8. https://doi.org/10.3390/nano8110942
12. Kauling AP, Seefeldt AT, Pisoni DP et al (2018) The worldwide graphene flake production. Adv Mater 30:1803784. https://doi.org/10.1002/adma.201803784
13. Li X, Cai W, An J et al (2009) Large-area synthesis of high-quality and uniform graphene films on copper foils. Science 324:1312–1314. https://doi.org/10.1126/science.1171245
14. Reina A, Son H, Jiao L et al (2008) Transferring and identification of single- and few-layer graphene on arbitrary substrates. J Phys Chem C 112:17741–17744. https://doi.org/10.1021/jp807380s
15. Plutnar J, Pumera M, Sofer Z (2018) The chemistry of CVD graphene. J Mater Chem C 6:6082–6101. https://doi.org/10.1039/c8tc00463c
16. Son J, Lee S, Kim SJ et al (2016) Hydrogenated monolayer graphene with reversible and tunable wide band gap and its field-effect transistor. Nat Commun 7:13261. https://doi.org/10.1038/ncomms13261
17. Jin Z, McNicholas TP, Shih CJ et al (2011) Click chemistry on solution-dispersed graphene and monolayer CVD graphene. Chem Mater 23:3362–3370. https://doi.org/10.1021/cm201131v
18. Bottari G, Herranz MÁ, Wibmer L et al (2017) Chemical functionalization and characterization of graphene-based materials. Chem Soc Rev 46:4464–4500. https://doi.org/10.1039/C7CS00229G
19. Kaplan A, Yuan Z, Benck JD et al (2017) Current and future directions in electron transfer chemistry of graphene. Chem Soc Rev 46:4530–4571. https://doi.org/10.1039/c7cs00181a
20. Greenwood J, Phan TH, Fujita Y et al (2015) Covalent modification of graphene and graphite using diazonium chemistry: tunable grafting and nanomanipulation. ACS Nano 9:5520–5535. https://doi.org/10.1021/acsnano.5b01580
21. Niyogi S, Bekyarova E, Itkis ME et al (2010) Spectroscopy of covalently functionalized graphene. Nano Lett 10:4061–4066. https://doi.org/10.1021/nl1021128
22. Malard LM, Pimenta MA, Dresselhaus G, Dresselhaus MS (2009) Raman spectroscopy in graphene. Phys Rep 473:51–87. https://doi.org/10.1016/j.physrep.2009.02.003
23. Jorio A, Dresselhaus MS, Riichiro Saito GD (2011) Raman spectroscopy in graphene related systems. Weinheim, Wiley-VCH Verlag GmbH
24. Cançado LG, Jorio A, Ferreira EHM et al (2011) Quantifying defects in graphene via Raman spectroscopy at different excitation energies. Nano Lett 11:3190–3196. https://doi.org/10.1021/nl201432g
25. Frank O, Kavan L, Kalbac M (2014) Carbon isotope labelling in graphene research. Nanoscale 6:6363–6370. https://doi.org/10.1039/C4NR01257G
26. Saito R, Jorio A, Souza Filho AG et al (2001) Probing phonon dispersion relations of graphite by double resonance Raman scattering. Phys Rev Lett 88:27401. https://doi.org/10.1103/PhysRevLett.88.027401
27. Eckmann A, Felten A, Mishchenko A et al (2012) Probing the nature of defects in graphene by Raman spectroscopy. Nano Lett 12:3925–3930. https://doi.org/10.1021/nl300901a
28. Lazzeri M, Mauri F (2006) Nonadiabatic Kohn anomaly in a doped graphene monolayer. Phys Rev Lett 97:266407. https://doi.org/10.1103/PhysRevLett.97.266407
29. Yan J, Zhang Y, Kim P, Pinczuk A (2007) Electric field effect tuning of Electron-phonon coupling in graphene. Phys Rev Lett 98:166802. https://doi.org/10.1103/PhysRevLett.98.166802

30. Das A, Pisana S, Chakraborty B et al (2008) Monitoring dopants by Raman scattering in an electrochemically top-gated graphene transistor. Nat Nanotechnol 3:210–215. https://doi.org/10.1038/nnano.2008.67
31. Verhagen T, Vales V, Frank O et al (2017) Temperature-induced strain release via rugae on the nanometer and micrometer scale in graphene monolayer. Carbon N Y 119:483–491. https://doi.org/10.1016/j.carbon.2017.04.041
32. Lee JE, Ahn G, Shim J et al (2012) Optical separation of mechanical strain from charge doping in graphene. Nat Commun 3:1024. https://doi.org/10.1038/ncomms2022
33. Verhagen TGA, Drogowska K, Kalbac M, Vejpravova J (2015) Temperature-induced strain and doping in monolayer and bilayer isotopically labeled graphene. Phys Rev B Condens Matter Mater Phys 92:125437. https://doi.org/10.1103/PhysRevB.92.125437
34. Kalbac M, Frank O, Kong J et al (2012) Large variations of the Raman signal in the spectra of twisted bilayer graphene on a BN substrate. J Phys Chem Lett 3:796–799. https://doi.org/10.1021/jz300176a
35. Jorio A, Cançado LG (2013) Raman spectroscopy of twisted bilayer graphene. Solid State Commun 175–176:3–12. https://doi.org/10.1016/j.ssc.2013.08.008
36. Fang W, Hsu AL, Caudillo R et al (2013) Rapid identification of stacking orientation in isotopically labeled chemical-vapor grown bilayer graphene by Raman spectroscopy. Nano Lett 13:1541–1548. https://doi.org/10.1021/nl304706j
37. Costa SD, Ek Weis J, Frank O, Kalbac M (2016) Effect of layer number and layer stacking registry on the formation and quantification of defects in graphene. Carbon N Y 98:592–598. https://doi.org/10.1016/j.carbon.2015.11.045
38. Terrones H, Lv R, Terrones M, Dresselhaus MS (2012) The role of defects and doping in 2D graphene sheets and 1D nanoribbons. Rep Prog Phys 75:62501. https://doi.org/10.1088/0034-4885/75/6/062501
39. Banhart F, Kotakoski J, Krasheninnikov AV (2011) Structural defects in graphene. ACS Nano 5:26–41. https://doi.org/10.1021/nn102598m
40. Frank O, Vejpravova J, Kavan L, Kalbac M (2013) Raman spectroscopy investigation of defect occurrence in graphene grown on copper single crystals. Phys Status Solidi Basic Res 250:2653. https://doi.org/10.1002/pssb.201300065
41. Jorio A, Lucchese MM, Stavale F et al (2010) Raman study of ion-induced defects in N-layer graphene. J Phys Condens Matter 22:334204. https://doi.org/10.1088/0953-8984/22/33/334204
42. Paulus GLC, Wang QH, Strano MS (2013) Covalent electron transfer chemistry of graphene with diazonium salts. Acc Chem Res 46:160–170. https://doi.org/10.1021/ar300119z
43. Wang QH, Jin Z, Kim KK et al (2012) Understanding and controlling the substrate effect on graphene electron-transfer chemistry via reactivity imprint lithography. Nat Chem 4:724–732. https://doi.org/10.1038/nchem.1421
44. Plšek J, Kovaříček P, Valeš V, Kalbáč M (2017) Tuning the reactivity of graphene by surface phase orientation. Chem Eur J 23:1839–1845. https://doi.org/10.1002/chem.201604311
45. Fan X, Nouchi R, Tanigaki K (2011) Effect of charge puddles and ripples on the chemical reactivity of single layer graphene supported by SiO_2/Si substrate. J Phys Chem C 115:12960–12964. https://doi.org/10.1021/jp202273a
46. Whitener KE (2018) Review article: hydrogenated graphene: a user's guide. J Vac Sci Technol A 36:05G401. https://doi.org/10.1116/1.5034433
47. Drogowska K, Kovaříček P, Kalbáč M (2017) Functionalization of hydrogenated chemical vapour deposition-grown graphene by on-surface chemical reactions. Chem Eur J 23:4073–4078. https://doi.org/10.1002/chem.201605385
48. Nair RR, Ren W, Jalil R et al (2010) Fluorographene: a two-dimensional counterpart of Teflon. Small 6:2877–2884. https://doi.org/10.1002/smll.201001555
49. Wang B, Wang J, Zhu J (2014) Fluorination of graphene: a spectroscopic and microscopic study. ACS Nano 8:1862–1870. https://doi.org/10.1021/nn406333f
50. Feng W, Long P, Feng Y, Li Y (2016) Two-dimensional fluorinated graphene: synthesis, structures, properties and applications. Adv Sci 3:1500413. https://doi.org/10.1002/advs.201500413

51. Kovaříček P, Bastl Z, Valeš V, Kalbac M (2016) Covalent reactions on chemical vapor deposition grown graphene studied by surface-enhanced Raman spectroscopy. Chem Eur J 22:5404–5408. https://doi.org/10.1002/chem.201504689
52. Kovaříček P, Vrkoslav V, Plšek J et al (2017) Extended characterization methods for covalent functionalization of graphene on copper. Carbon N Y 118:200–207. https://doi.org/10.1016/j.carbon.2017.03.020
53. Robinson JT, Burgess JS, Junkermeier CE et al (2010) Properties of fluorinated graphene films. Nano Lett 10:3001–3005. https://doi.org/10.1021/nl101437p
54. Ek Weis J, Costa SD, Frank O et al (2015) Fluorination of isotopically labeled turbostratic and bernal stacked bilayer graphene. Chem Eur J 21:1081–1087. https://doi.org/10.1002/chem.201404813
55. Wang YY, Ni ZH, Shen ZX et al (2008) Interference enhancement of Raman signal of graphene. Appl Phys Lett 92:43121. https://doi.org/10.1063/1.2838745
56. Stiles PL, Dieringer JA, Shah NC, Van Duyne RP (2008) Surface-enhanced Raman spectroscopy SERS: surface-enhanced Raman spectroscopy Raman scattering: inelastic scattering of a photon from a molecule in which the frequency change precisely matches the difference in vibrational energy levels. Annu Rev Anal Chem 1:601–626. https://doi.org/10.1146/annurev.anchem.1.031207.112814
57. Schedin F, Lidorikis E, Lombardo A et al (2010) Surface-enhanced Raman spectroscopy of graphene. ACS Nano 4:5617–5626. https://doi.org/10.1021/nn1010842
58. Lee J, Shim S, Kim B, Shin HS (2011) Surface-enhanced Raman scattering of single-and few-layer graphene by the deposition of gold nanoparticles. Chem Eur J 17:2381–2387. https://doi.org/10.1002/chem.201002027
59. Hu Y, López-Lorente ÁI, Mizaikoff B (2019) Graphene-based surface enhanced vibrational spectroscopy: recent developments, challenges, and applications. ACS Photonics 6:2182. https://doi.org/10.1021/acsphotonics.9b00645
60. Zhu X, Shi L, Schmidt MS et al (2013) Enhanced light-matter interactions in graphene-covered gold nanovoid arrays. Nano Lett 13:4690–4696. https://doi.org/10.1021/nl402120t
61. Weis JE, Vejpravova J, Verhagen T et al (2018) Surface-enhanced Raman spectra on graphene. J Raman Spectrosc 49:168–173. https://doi.org/10.1002/jrs.5228
62. Ghamsari BG, Olivieri A, Variola F, Berini P (2015) Frequency pulling and line-shape broadening in graphene Raman spectra by resonant Stokes surface plasmon polaritons. Phys Rev B 91:201408. https://doi.org/10.1103/PhysRevB.91.201408
63. Economou EN (1969) Surface plasmons in thin films. Phys Rev 182:539–554. https://doi.org/10.1103/PhysRev.182.539
64. Rai VN, Srivastava AK, Mukherjee C, Deb SK (2012) Surface enhanced absorption and transmission from dye coated gold nanoparticles in thin films. Appl Opt 51:2606. https://doi.org/10.1364/AO.51.002606
65. Sutrová V, Šloufová I, Mojzeš P et al (2018) Excitation wavelength dependence of combined surface- and graphene-enhanced Raman scattering experienced by free-base phthalocyanine localized on single-layer graphene-covered Ag nanoparticle arrays. J Phys Chem C 122:20850–20860. https://doi.org/10.1021/acs.jpcc.8b06218
66. Kalbac M, Vales V, Vejpravova J (2014) The effect of a thin gold layer on graphene: a Raman spectroscopy study. RSC Adv 4:60929. https://doi.org/10.1039/c4ra11270a
67. Nan H, Chen Z, Jiang J et al (2018) The effect of graphene on surface plasmon resonance of metal nanoparticles. Phys Chem Chem Phys 20:25078–25084. https://doi.org/10.1039/c8cp03293a

Chapter 2
Applications of Graphite Materials in the Field of Electromagnetic Compatibility

Octavian Baltag and Georgiana Rosu

2.1 Introduction

The theoretical foundation (1864) of the existence of electromagnetic waves [1] was formulated by James Clerk Maxwell and their generation and detection in 1886 was achieved by Heinrich Hertz [2]. After that began the era of electromagnetic communications. The first transmission—a reception system experienced (1894) in the laboratory by Oliver Lodge [3], was followed by the first wireless telegraphic communication system (at a distance of 2 km) achieved by Gugliemo Marconi [4] in 1895, culminating with the visionary and spectacular system of Nikola Tesla but also dangerous and without a practical completion, on the wireless transmission of electricity [5]. The US military has awakened potential with great prospects for long-distance communications, beyond the visible horizon. In 1899, wireless telegraphy was used in US military vessels, and the system was to be assimilated into civilian communications. Because the broadcast systems were made with discharge generators, that used the same frequency spectrum, receiving messages was difficult and unintelligible due to interference. At this point, there arose the first serious problem regarding what nowadays is called Electro Magnetic Compatibility—EMC. This was the "Radio Frequency Interference—RFI."

This was the beginning of the formulation of EMC regulations regarding the use of EM waves in communication. Thus, the need arose to assign different areas of frequency to telegraphy stations as well as to new radio stations belonging to the army and the private sector. World War II led not only to the development of electronic and radar communication, but also to the emergence of severe problems,

O. Baltag (✉)
"Grigore T. Popa" University of Medicine and Pharmacy of Iasi, Iasi, Romania
e-mail: octavian.baltag@bioinginerie.ro

G. Rosu
"Ferdinand I" Military Technical Academy, Bucharest, Romania

C. Miron et al. (eds.), *Carbon-Related Materials*,
https://doi.org/10.1007/978-3-030-44230-9_2

particularly in the security of military communications—air, naval, and land that motivated the US Navy to develop a first RFI standard after the war. Another more serious problem with disastrous effects emerged as a result of nuclear experiments: Nuclear Electro Magnetic Pulse—NEMP.

After this period, over time, the level and spectrum of EM fields became more and more complex, as a result of the use of electricity in rail transport, the increase of the density of electrical networks, but also the increasing use of domestic equipment, determining the emergence of physiological and behavioral medical symptoms specific to the activities performed in the presence of these fields.

In order to protect against these fields, the electromagnetic shielding properties of different material types were studied. The first materials used in the field of electromagnetic compatibility were metallic ones—steel, copper, silver, gold, with very good shielding properties but with the disadvantage of a high mass density, which limits their applicability. To avoid this disadvantage, carbon-based materials were adopted. Graphite is one of the two allotropic states of carbon, in which the atomic network—also called the stratified network. This is made up of parallel planes composed of carbon atoms arranged in a regular hexagonal structure. Unlike metals, graphite has an extremely low chemical reactivity.

2.2 Current Trends

International regulations and standards have set clear limits on the spectrum and intensity of electric, magnetic, and electromagnetic fields (EMFs), depending on the human activity area. In order to reduce the health effects of these fields, according to the professional or domestic activities, it was necessary to divide the electromagnetic protection materials into two classes: class I of the materials for professional use and class II of the materials for general use. The research in this field is permanently directed towards the elaboration of new regulations regarding the sources of EM pollution and the implementation of new techniques and products of human protection.

Currently, electromagnetic shielding is obtained by using textile composite structures and textile products with shielding properties [6]. For this purpose, several types of textile materials (textile composite structures, knits, and fabrics) have been developed, characterized, and analyzed. These materials have different parameters and structures and contain several categories of yarns (amorphous magnetic yarns, copper yarns, silver yarns, and stainless steel wires) as well as embedded graphite powder, which exhibits electromagnetic properties specific to EM fields. Also, metallic fibers have been used to make yarns for fabrics and knits, and it was subsequently turned to the manufacturing of metallic coated textiles through various methods—chemical deposition, sputtering.

A field of special interest is represented by the applications of graphene for shielding in the terahertz frequency range and to avoid electromagnetic interference and is found in: [7–12]. Likewise, devices based on metamaterials made up of graphene with electronic control of the parameters for applications in active filtration and tunable frequencies in the terahertz range have been studied in [13]. Specific technologies for the preparation of graphene-like structures are presented in numerous patents [14–19]. In [20] the absorbing properties of graphene devices, using plasmonic resonance, with possible applications in modulators, polarizers, adjustable filters, etc. are studied in the 0.4 THz frequency range. Another relatively inexpensive technology refers to the use of screen printing methods for the realization of devices with shielding properties based on materials incorporating carbon [21].

Other research quotes the use of conductive particles of copper, silver, and carbon applied to the textile material by appropriate techniques in order to make conductive textile materials. In recent years, research has focused on the use of conductive polymers, such as polyacetylene, polypyrrole, and polyaniline, applied to textile materials. Therefore, the production of these composites has shown satisfactory results for the protection of the human body. The frequency range of particular interest also includes the microwave spectrum used in GSM communication systems and wireless data transmission.

Graphite and carbon black applications in EM shielding products are: the shielding gaskets, housing, grounding clips, conductive foam, EMI rubber, EMI & RFID absorbers, personal protection, shielding pouches, etc. The advantages of using graphite are: high electrical conductivity, lower mass density compared to metals, corrosion resistance in hostile environments, mechanical flexibility, and easy processing. Of all the basic forms of graphite, graphene has the highest potential for its use in EM shielding, besides applications in nanoelectronics. Its use in structures can be applied in the form of weather-resistant paint—water, snow, temperature, corrosion, etc., with shielding paint properties; this ensures a low cost of application and protection against electrostatic discharges.

2.3 Electromagnetic Shielding Mechanism

Material conductivity is directly related to the shielding properties and the ability to conduct the electric current. Graphite has a lower electrical conductivity of about three orders of magnitude compared to the ones of metals. Although graphite has properties far superior to polymeric materials, natural textiles, artificial resins, in the form of fibers or granules—nanocomposites, their presence diminishes the conductive properties depending on the concentration of the used ingredients and their ratio to the amount of graphite [22, 23]. These properties also depend on the shape of the final composite material: tiles or wires. Generally, graphite is introduced into composite structures that do not have electrical conductivity in the form of powder or wires; the wires are preferred due to their higher mechanical resistance, particularly against the forces applied in the direction of the fiber. Regarding electrical conductivity, it is also higher in the direction of graphite fibers.

For these reasons, the electrical and electromagnetic characterization of graphite composite materials must comply with certain specific procedures. The sample subjected to the characterization must be made up of standard geometry so that the following quantities can be measured: volume resistivity and surface resistivity, both in the direction of the fibers and perpendicular to them. At the individual measurement of fibers, their ends should be covered with conductive paint. At high frequencies, the sample is characterized by intrinsic impedance, which is dependent on permittivity, permeability, frequency, conductivity, and the sample geometry, according to the relationship:

$$Z = \sqrt{\frac{j\omega\mu}{\sigma + j\omega\varepsilon}}. \tag{2.1}$$

This impedance is generally relative to the air impedance $Z_{air} = 377\ \Omega$, or to the impedance of a metallic material. Another essential parameter in electromagnetic shielding is related to the skin depth, which refers to the penetration depth of eddy currents at a certain frequency, their decrease being dependent on the factor $1/e$ to the current value at the surface of the material.

The EMF interaction with matter is described by complex mechanisms that depend on a number of factors: wave frequency, energy density, electromagnetic properties of the structure subjected to EMF, and the nature of the propagating environment. The shielding phenomena can be divided into:

- Attenuation due to electromagnetic energy absorption into the material
- Attenuation through external reflections
- Mitigation through successive internal reflections in its protective environment

Depending on the electrical and magnetic components of the EMF, attenuation is achieved through:

- Electrostatic attenuation, which is achieved by the interposition of components with high electrical conductivity. This ensures an equipotential surface by grounding, which has a null electric field inside.
- Magnetic attenuation, which is achieved by interposing components with high magnetic permeability that ensures refraction of the magnetic field.

The shielding factor F is used for the assessment of the shielding effect and is represented as a ratio of the electric (E)/magnetic (H) components inside the protected area, relative to the values obtained in the absence of the screen:

$$F = \frac{H_{in}}{H_{ex}} \angle 1. \tag{2.2}$$

If we consider a non-ferromagnetic wall—made of graphite, for example, with high electrical conductivity σ, pulsation ω, and permittivity ε, for this wall the condition $\sigma \gg \omega\varepsilon$ is met. In these conditions, the intrinsic impedance is given by the relation:

$$Z_{in} = \sqrt{\frac{j\omega\mu}{\sigma + j\omega\varepsilon}} = \sqrt{\frac{j\omega\mu}{\sigma}}\sqrt{\frac{\omega\mu}{2\sigma}}(1+j) = \sqrt{\frac{\omega\mu}{\sigma}}e^{j\frac{\pi}{4}}. \tag{2.3}$$

Equation (2.3) describes an inductive impedance:

$$Z_{in} = R + j\omega L. \tag{2.4}$$

From the above equations, the two components of the shield's intrinsic impedance can be expressed:

$$R = \frac{1}{\sigma\delta},$$
$$L = \frac{\mu\delta}{2}. \tag{2.5}$$

Here δ denotes the field penetration depth inside the shielding material. Therefore, the intrinsic impedance yields the expression:

$$Z_{in} = \frac{1}{\sigma\delta} + j\omega\frac{\mu\delta}{2}. \tag{2.6}$$

Due to the fact that the *E* and *H* components are generally harmonic, and because of the phase difference between the two inner and outer components, the shielding factor is expressed as a complex quantity. If the EMF is a plane wave, the *E*/*H* ratio is constant and equal to the impedance Z_0 of the propagation medium, and the attenuation factor can also be calculated by the ratio of the electrical components. In practice, it is preferable to use the so-called screening/shielding effectiveness:

$$S = 10\log\left|\frac{P_{in}}{P_{ex}}\right| 20\log\left|\frac{E_{in}}{E_{ex}}\right| [\text{dB}]. \tag{2.7}$$

When an electromagnetic wave encounters a surface with high electrical conductivity, the phenomena taking place are complex: all three types of losses occur, each of them with a different weight depending on the size of the screen reported to the wavelength and its electromagnetic properties; ultimately, all of these losses will contribute to the shielding effectiveness. Shielding factor *F* and screening efficiency *S* depend on some factors: the place where the measurement is performed, the field component being measured, the field incidence, the type of polarization as well as the nature and geometry of the screen [24].

The protective measures are taken according to the target or the environment that must be protected against EMF. Related to this fact, the distance between the source and the target, relative to the field's wavelength, plays an important part. Thus, if the disturbance source is an antenna located far away from the target, compared to the wavelength, the propagation of the electromagnetic wave takes place in the domain called the far field, but if the source is, for example, inside a box, the propagation occurs in an area called the near field region. These situations determine specific propagation conditions and particularities of the protection procedures.

The radiation of an antenna—the source of the EMF, can be divided into three main areas (see Fig. 2.1), depending on the power density, distance, and especially the characteristics of the E and H field components. It should be noted that in the space between the physical structure of the antenna and the radiation zone, the transformation of the E and H field configuration is continuous, although there are significant differences between the power density and the field configuration.

The Rayleigh area is called the non-radiation, inductive, very-near field area, where the reactive energy exchange occurs between the antenna and the surrounding environment. In this area, the distribution of electric and magnetic fields is highly dependent on the size and geometry of the antenna and is strongly variable with distance. The E and H components are not perpendicular and have a relatively high intensity, and for high power field it is a dangerous area (compared to the radiation zone) due to the strong interaction of the electrical component with the biological media. The Rayleigh distance extends from the antenna to $r = L^2/2\lambda$ and includes the induction zone, with the highest energy density. The field equations in this region include imaginary terms which show the presence of reactive energy, both reactive components of the field—electric and magnetic, being predominant. It can be said that this is the region where the energy is stored.

The near field radiation area is called *the Fresnel area* and is delimited by the distances $L^2/2\lambda \leq r \leq 2\,L^2/\lambda$ (for $L = \lambda$, it results $\lambda^2/2 \leq r \leq 2\lambda$). In this area, the E and H components are not perpendicular, and their vectors have a variable angle

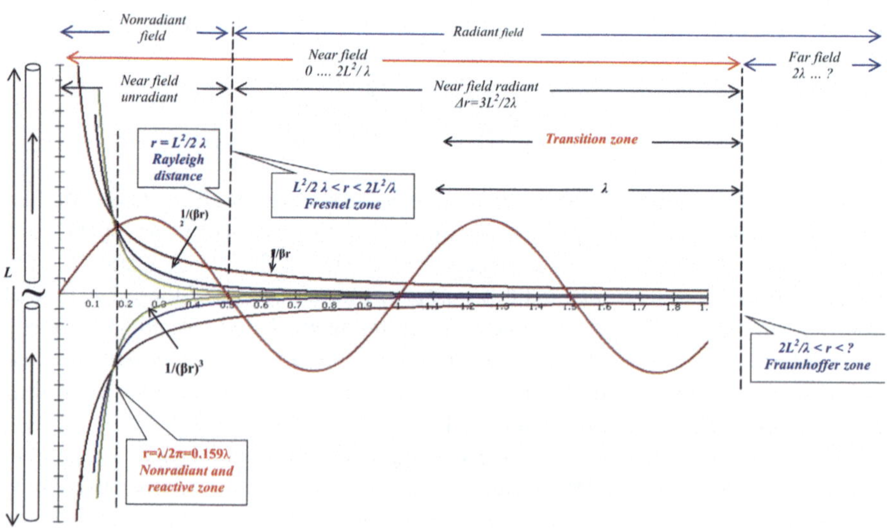

Fig. 2.1 Near and far field regions, for the case of antenna length $L = \lambda$ [25]

with respect to the antenna, tending to become perpendicular as the field approaches the Fraunhofer area. For a large antenna, compared to the field's wavelength, the upper limit is $2L^2/\lambda$. Near this limit and after that, the radiant field is predominant, and the angular distribution of the power density depends on the distance from the source. This is considered by some authors as a transition zone that stretches (1–2)λ from the antenna to the Fraunhofer region. It is an area with intermediate effects, where the near field effects are reduced, and instead the far field effects become more prominent.

The far field radiation area—*the Fraunhofer area*, has the lower limit of $2L^2/\lambda$ and extends to infinity; in this region, the E and H components become perpendicular, the spherical wave is transformed into a plane wave, and the radiation directions become parallel. There is an identical distribution of electromagnetic energy between the E and H components. The distance from which the wave can be considered the plane wave with a relatively good approximation, is determined by the plane in which the phase error is accepted; for the lower limit of $2L^2/\lambda$, this error is 22.5°, or $\lambda/16$ (the Fraunhofer condition).

The region of interest depends on the area intended for protection: if there is the case of shielding equipment or biological matter from outside, then the protected sample lies in the near field area—Rayleigh or Fresnel, depending on the shielded volume. On the other hand, if the source of EMF is located inside and the protected area lies outside the shield, then the field region of interest is the Fraunhofer area.

2.4 Electromagnetic Wave Propagation in Non-magnetic Conductive Media

Regarding the EMF propagation when encountering an obstacle made of a non-magnetic material, such as a structure containing graphite, the wave energy losses through absorption must be taken into account along with the losses through reflection.

2.4.1 Absorption Loss

In order to calculate the absorption loss in the case of graphite, one can consider a super unitary ratio between the conduction current and the displacement current in the respective environment. For a plane, if a harmonic electromagnetic wave penetrating the graphite wall in the z direction, the equation of the field $H_y(z, t)$ is:

$$H_y(z,t) = H_0 e^{-\frac{z}{\delta}} \left(\cos \omega t - \frac{z}{\delta} \right), \tag{2.8}$$

where z/δ is the phase shift at a distance z from the input. At distance z, the amplitude of the field is:

$$H = H_0 e^{-\frac{z}{\delta}}. \tag{2.9}$$

Therefore, the field decreases exponentially with distance, and for $z = \delta$ the decrease in amplitude will be equal to $e = 2.71$. This is illustrated in Fig. 2.2.

The relationship can be expressed as:

$$A = 20\log\left|\frac{H_0}{H_z}\right| = 20\frac{z}{\delta}\log e = 8.69\frac{d}{\delta}, \tag{2.10}$$

where d denotes the depth of field's penetration inside the considered material. The penetration depth depends on the material properties and is defined by Eq. (2.11):

$$\delta = \sqrt{\frac{1}{\pi f \sigma \mu}}. \tag{2.11}$$

This relation indicates a dependence on the incident field frequency of both the resistive R and the inductive L components; also, the inductive component is inversely proportional to the material conductivity σ.

2.4.2 *Reflection Loss*

Consider a plane wave (called a direct wave) with normal incidence on the separation surface between a medium of impedance Z_1 and the graphite medium—with impedance Z_2. Taking into account that the two environments are analogous to a transmission line, the same calculation formalism can be applied. Thus, the direct wave—the incident wave, will be divided at the separation surface into two waves, a transmitted wave that will continue the propagation in the new environment and a wave reflected backwards, as illustrated in Fig. 2.3.

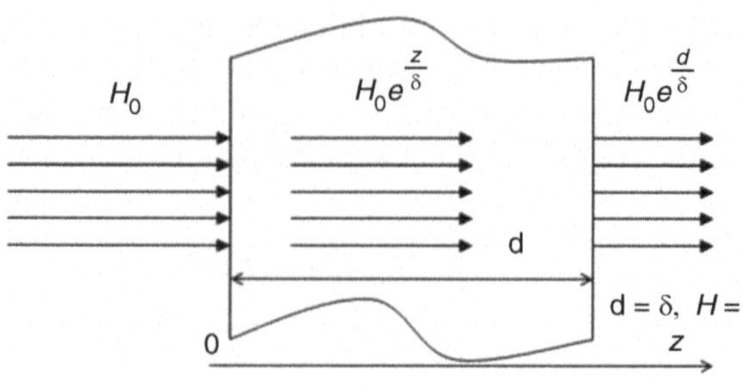

Fig. 2.2 Field attenuation when passing through a wall/shield

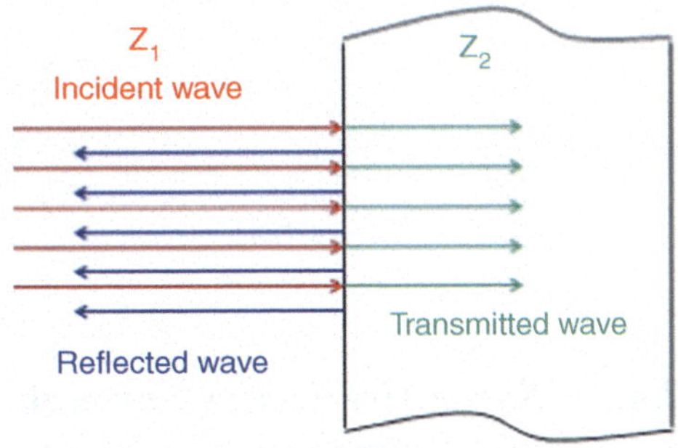

Fig. 2.3 Shielding through the reflection mechanism

Since the wave incidence is perpendicular to the surface, it turns out that all the *E* and *H* components of the direct and reflected electromagnetic fields are tangential to the surface. These electric and magnetic field components are continuous and are summed up at the surface, which simplifies the calculation. Analogous to the transmission line, the medium Z_1 can be considered as a transmission line having a load equal to impedance Z_2 of the second environment (the graphite). Thus, the reflection coefficients ρ_E and ρ_H for the electric and magnetic components are defined as:

$$\rho_E = \frac{E_r}{E_d}, \quad \rho_H = \frac{H_r}{H_d}. \tag{2.12}$$

where $\rho_E = -\rho_H$. These coefficients can be expressed by means of the impedances of the two environments:

$$\rho_E = \frac{Z_2 - Z_1}{Z_2 + Z_1}, \quad \rho_H = \frac{Z_1 - Z_2}{Z_1 + Z_2}. \tag{2.13}$$

The transmitted wave can be expressed through the transmission coefficient, equal to the ratio of the two electric fields—the transmitted one E_t and the direct one E_d, respectively:

$$\tau_E = \frac{E_t}{E_d}. \tag{2.14}$$

The transmission coefficient for both field components (*E* and *H*, respectively) can also be expressed by means of the media impedances:

$$\tau_E = \frac{2Z_2}{Z_1 + Z_2},$$
$$\tau_H = \frac{2Z_1}{Z_1 + Z_2}. \tag{2.15}$$

2.4.3 Successive Losses Through Multiple Reflection and Refraction

Losses due to multiple reflections are less significant when the wall thickness is greater than the penetration depth because the wave attenuates rapidly. In the case of thin walls, significant multiple reflections can emerge.

If the incident wave is not perpendicular to the surface, then these losses increase, reaching a maximum for directions parallel with the separation surface; the wave is deflected outwards. The effect of multiple reflections is illustrated in Fig. 2.4, where the term I_A denotes the absorption factor of the reflected wave inside the screen. After a number of successive reflections and absorptions—with the attenuation following the law of geometric progression, a part of the wave is retransmitted to the environment where it comes from, and another component passes into the protected environment, which reduces the effect of the screen. Apart from multiple reflections inside the material and outside it, the refraction phenomenon also appears—with a damped variation.

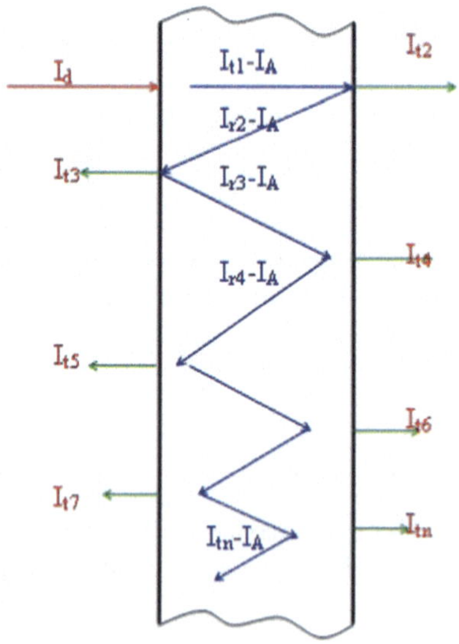

Fig. 2.4 Attenuation through successive reflections and refractions. I_d—incident wave, I_t (blue)—the wave transmitted inside the shield, I_A—the absorption loss inside the shield, I_t (green, left)—the reflected wave, I_t (green, right)—the transmitted wave after passing through the screen

2.4.4 The Shielding Effectiveness General Equation

The manifestation of all cumulative phenomena related to the interaction of the electromagnetic field with the barrier used for shielding is characterized by the relation:

$$SE = A + R + RM. \tag{2.16}$$

Here the following notations were used: SE represents the total attenuation or the shielding effectiveness, A represents the wave absorption attenuation, R—the wave reflection attenuation, and RM—the attenuation through multiple reflections inside the screen. All these phenomena are illustrated in Fig. 2.5. The shielding performances of the screen can be determined by direct measurements of the transmitted or reflected components, whereas the multiple reflections result from the calculation.

2.5 Numerical Analysis of Carbon Powder Electromagnetic Shields

In this paragraph, the screen performances of different types of screens were analyzed, using dedicated software—Ansys HFSS (High Frequency Structure Simulator) based on the finite element method. In all the analyzed situations, a relatively small area was adopted, but large enough so that it would not influence the propagation mode of the wave. The discretization network is adapted from one iteration to the next, with a number of 20 adaptive passes being imposed. The program will create a tetrahedron mesh based on the geometry of the model, which it

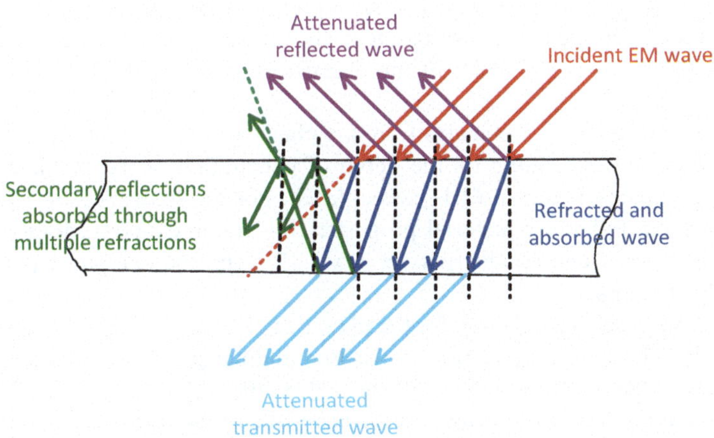

Fig. 2.5 The entire shielding mechanism [25]

will adapt from iteration to the next so that the tetrahedron's dimensions are smaller than a quarter of the wavelength corresponding to the input frequency—for example, for the frequency of 1 GHz the length the wave is 30 cm. The solution has a maximum deviation of the error of parameter *S*, Delta *S* = 1%, representing the difference between the field values obtained for two successive iterations.

In the first instance (Sect. 2.5.1), the shielding properties of graphite plates of different thicknesses expressed according to the depth of penetration of the electromagnetic field in graphite were studied. For this purpose, the frequency of the incident wave of 1 GHz (wavelength of 30 cm) was used, and the thickness of the graphite plate was varied.

Then the effectiveness of the screen was analyzed for a network made of graphite strips with a thickness of 0.1 mm, organized in two configurations: the strips arranged parallel and equidistant in one direction, and then the strips were disposed in two directions, thus constituting a grid structure.

The purpose of the graphite strip structure analysis—in Sect. 2.5.2, was to evaluate the screen's effectiveness by means of the aperture size (the distance between the strips for the first configuration and the mesh size in the second configuration, respectively) relative to the wavelength of the incident field.

In Sect. 2.5.3, the ability of the graphite screen to shield, depending on the polarization of the incident wave, was investigated. For this scope, two separate incident waves with the same frequency, 1 GHz, and different polarization—vertical and horizontal polarization, respectively, were used.

2.5.1 The Analysis of Shielding Effectiveness Depending on the Width of the Screen

In order to study the shielding effectiveness in relation to the thickness of the graphite screen, a plane wave with 1 V/m amplitude, 1 GHz frequency, and vertical polarization was applied as an excitation on the left side of the domain. The propagation of the wave in the field, in the absence of any screen, is represented in Fig. 2.6.

Then, in the center of the analysis area, a plane-parallel graphite screen, with a variable thickness depending on the depth of penetration of the incident field in graphite, calculated with Eq. (2.11) was placed. For the frequency of 1 GHz and considering the following material properties for graphite—permeability $\mu_r \approx 1$, permittivity $\varepsilon_r = 1$, and conductivity $\sigma = 5 * 10^2$ S/m, the computed penetration depth is $\delta = 60.2$ μm.

Thus, three cases were chosen for the plate's thickness: $0.1 * \delta$, $1 * \delta$, and $10 * \delta$, respectively. The field propagation in the longitudinal plane presents a reflection due to the contact with the screen, illustrated on the left side of Fig. 2.7 and the attenuation of the transmitted wave on the right side of the figure.

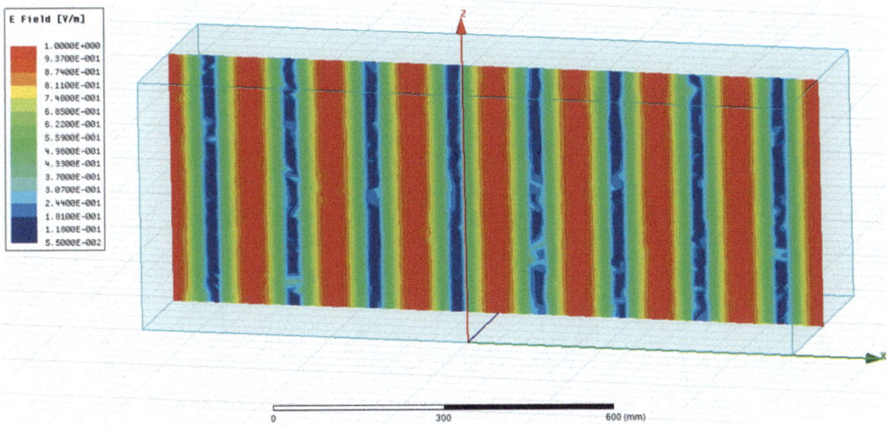

Fig. 2.6 Wave propagation inside the domain

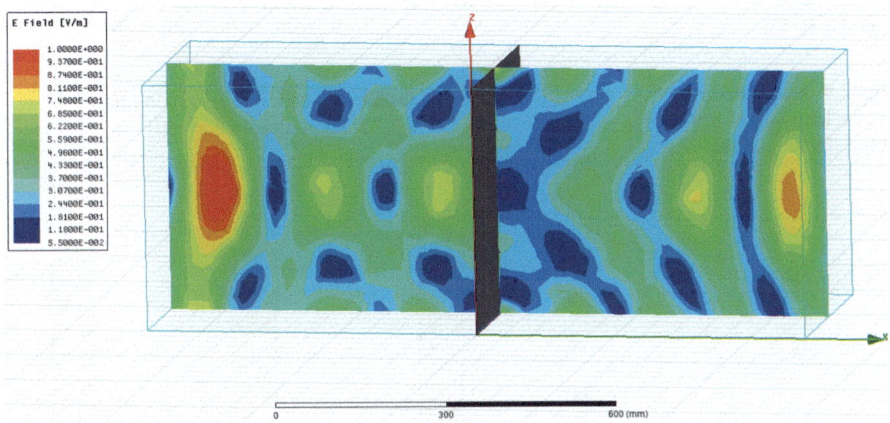

Fig. 2.7 Field propagation in the presence of the shield (width = 1 * δ)

In Fig. 2.8, the electric field values are given on a central longitudinal axis of the analysis domain, for all the analyzed cases: the field in the absence of the screen and the field in the presence of the screen with the three chosen thicknesses, respectively. In the presence of a screen, the attenuation of the transmitted field is noted.

In order to highlight the attenuation coefficients, the value of the reflected field and the field transmitted after passing through the screen are calculated in dB (dBV/m), in reference to the amplitude of the incident field. The resulting values are illustrated centrally in Fig. 2.9. The screen performance improvement as the thickness increases is obvious.

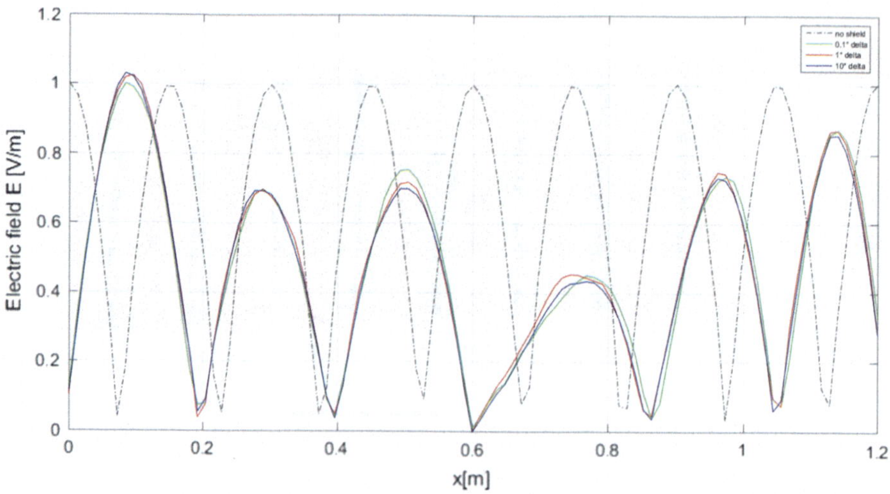

Fig. 2.8 Electric field strength (V/m) along the central line. Cases: black dotted line—without any shield, green line—shield width = 0.1 * δ, red line—shield width = 1 * δ, blue line—shield width = 10 * δ

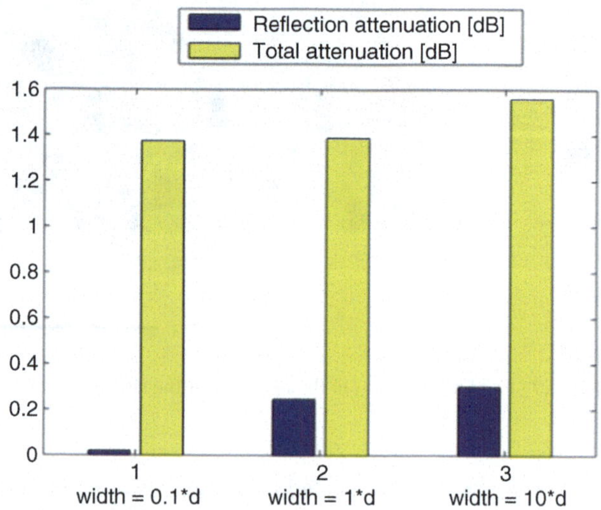

Fig. 2.9 Attenuation coefficients (dB) for the three shield widths

2.5.2 *The Analysis of the Shielding Effectiveness Depending on the Aperture Size*

When it comes to the assessment of the shielding effectiveness, an important criterion is the size of the aperture in relation to the wavelength of the incident field. For this purpose, a screen with dimensions similar to those of the plane-parallel screen in Sect. 2.5.1, but made up of parallel and equidistant strips was considered. Thus,

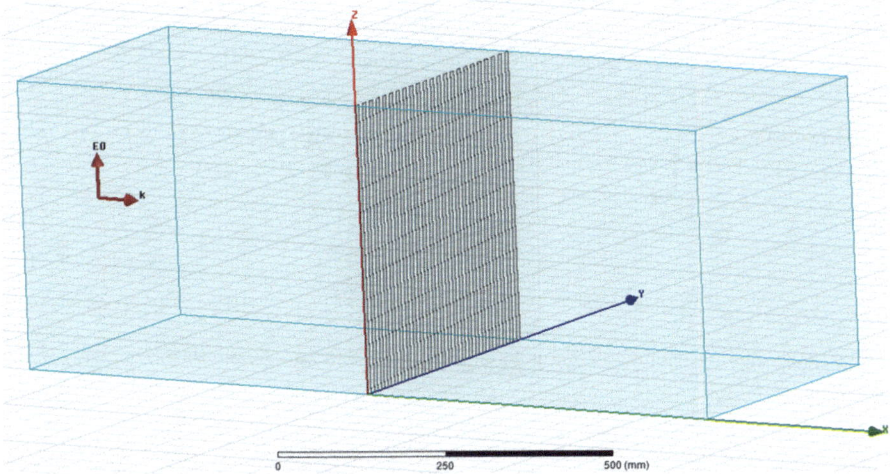

Fig. 2.10 Shield made of parallel graphite strips

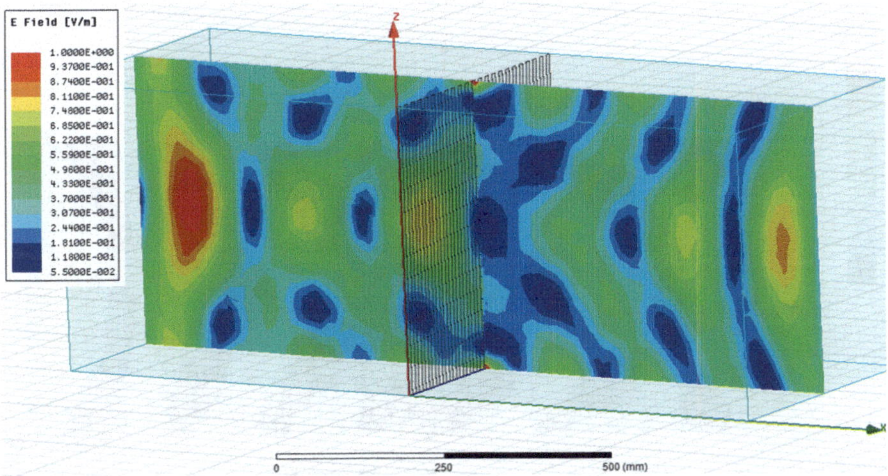

Fig. 2.11 Electric field distribution in longitudinal plan for incident field of 1 GHz

if the parallel plane screen in the previous paragraph was 450 mm wide, 450 mm high, and a variable thickness, in this case the screen consists of 23 strips 10 mm wide, 450 mm high, and 0.1 mm thick, arranged parallel and equidistantly at 10 mm—as shown in Fig. 2.10.

The incident plane wave has the same parameters as in paragraph 5.1: amplitude 1 V/m, frequency 1 GHz, and vertical polarization. Thus, for this frequency, the field penetration depth in graphite is $\delta = 60.2$ μm and the screen's width (of 0.1 mm) relative to the field wavelength $\lambda = 1.615 * \delta$. Figure 2.11 illustrates the field strength distribution in the longitudinal plane.

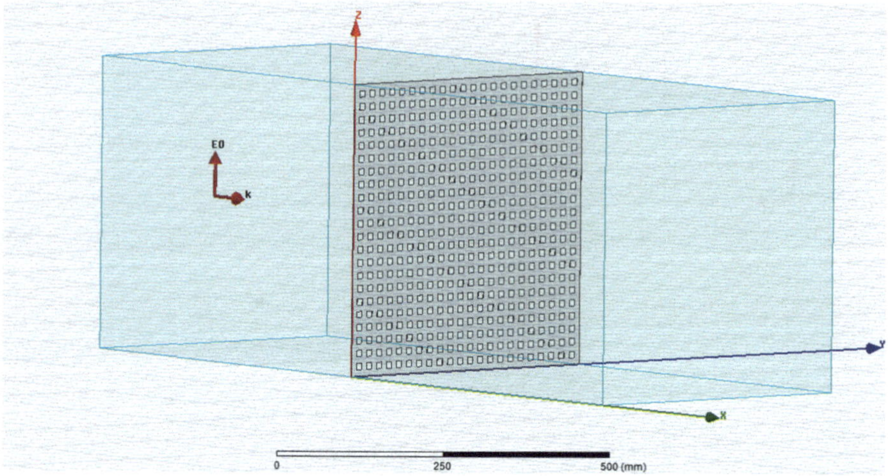

Fig. 2.12 Graphite grid screen made of perpendicular strips

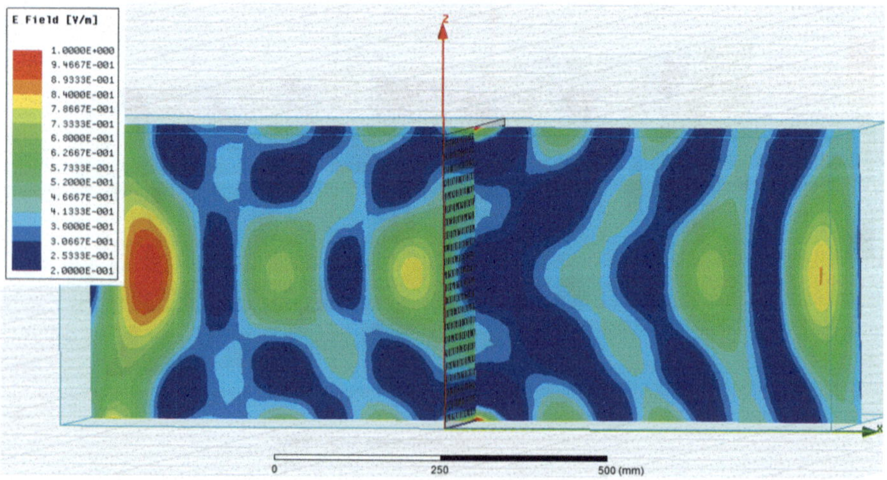

Fig. 2.13 Longitudinal field distribution for graphite grid at 1 GHz

Next, the effect of reducing the aperture size by creating a screen in the form of a grid structure made of graphite strips arranged in two perpendicular directions, as seen in Fig. 2.12 is studied. The strips have the same size as in the previous case, but the size of the screen aperture is significantly reduced.

The effect of this aperture size reduction on the propagation of the incident field is shown in Fig. 2.13. In order to accentuate this aspect, Fig. 2.14 shows the values of the electric field strength on the central longitudinal axis of the domain, for the two types of screens, at frequency of 1 GHz. The degree of optical trans-

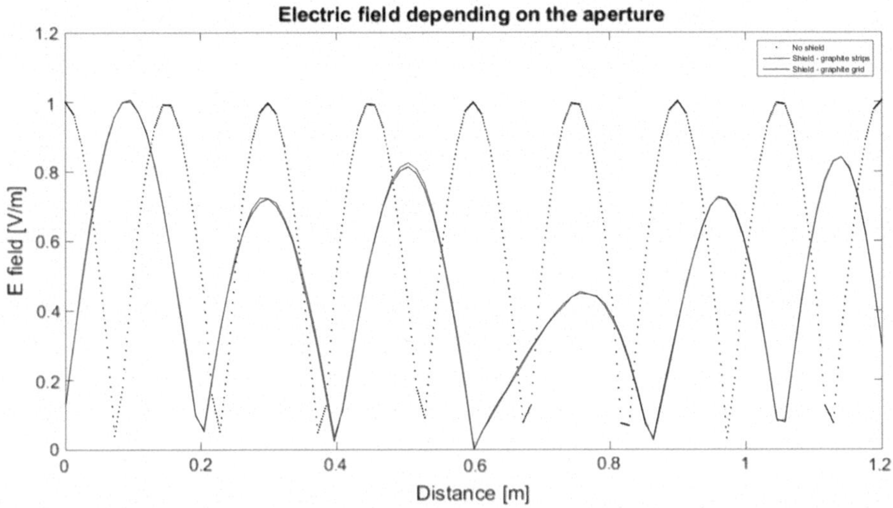

Fig. 2.14 Electric field strength (V/m) along the central line for graphite strips and graphite grid shields, relative to the situation when there is no shield

parency of the two screens, defined as the ratio between the transparent area for the field passage and the total area, is 49% for the parallel bands and 23.9% for the grid, respectively.

2.5.3 The Analysis of the Shielding Effectiveness Depending on the Incident Wave Polarization

Another significant aspect in electromagnetic shielding is the orientation of the aperture relative to the polarization of the incident wave. In order to highlight this aspect, we consider that the screen is made of parallel bands identical to the one in the previous paragraph, and we analyze the capacity of shielding a plane wave with vertical and horizontal polarization. The results obtained from the two simulations are illustrated in Fig. 2.15.

A higher shielding efficiency is observed in the first case because the incident wave has the same orientation as the direction of the graphite strip. In this case, the bands in the screen's structure act as a short circuit, inducing oscillations of the charge carriers in the screen's structure, in the direction of the wave polarization. These oscillations represent a secondary source that emits energy in all directions. Thus, much of the energy of the incident wave is re-radiated (reflected) in the environment from where it originated, and another part of the energy is allowed to pass into the environment on the right side of the screen.

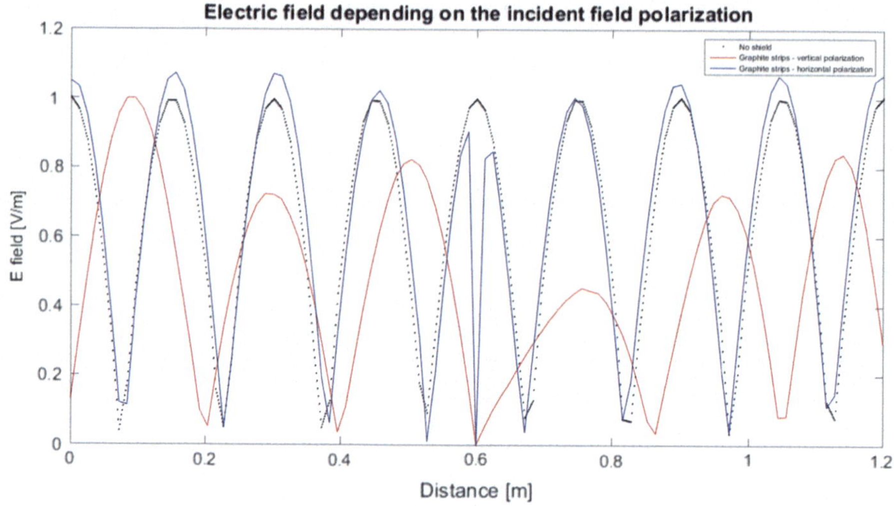

Fig. 2.15 The electric field strength on the longitudinal axis for vertical polarization (red line) and horizontal polarization (blue line) on the incident wave

2.6 Experimental Research on Selected Screen Configuration

In this chapter, the results obtained experimentally regarding the shielding ability of some manufactured screens and a carbon-impregnated fabric are presented.

2.6.1 Experimental Analysis of Shielding Effectiveness Depending on the Size of the Aperture and Incident Field Polarization

In the first stage, a set of graphite screens were manufactured, corresponding to those simulated in the previous chapter. Thus, the first screen consists of parallel strips, and the second screen is made up of a mesh of orthogonal strips. The graphite strips have a thickness of 0.1 mm, a width of 10 mm, and are placed at a distance of 10 mm, on a 60 cm × 60 cm plexiglas support. The degree of optical transparency of the two screens is approximately equal to that of the simulation screens—50% for the parallel bands and 21.6% for the network. The two constructed screens are illustrated in Fig. 2.16.

Fig. 2.16 The manufactured graphite shields: left—the parallel strips, right—the orthogonal grid

2.6.1.1 Experimental Determination of the Shielding Effectiveness

The shielding efficiency was measured in a semi-anechoic chamber, in which a cube of galvanized steel was mounted, and one of the sides of the cube was represented by the studied screen. A transmitting (Tx) antenna was mounted outside the cube and the receiving (Rx) antenna was placed inside the cube, thus eliminating any outer influence. The measurement distance between the two antennas was 2 m. The experimental setup is schematically illustrated in Fig. 2.17.

The field level measurements were made in two frequency bands: 100 MHz-1 GHz and 5–6 GHz, respectively. The results are presented comparatively for the two screens in Figs. 2.18 and 2.19, respectively. Although the attenuation values are low, a pattern of attenuation increase is observed for all the screens, in both bands studied, similar to a resonance frequency, due to the arrangement of the graphite strips.

2.6.1.2 Experimental Determination of the Reflection and Transmission Coefficient

In order to test the reflection and transmission coefficients of the aforementioned shields, two experimental setups were employed. In both setups, two directive antennas and a vector network analyzer (VNA) were employed, with both antennas connected with low-loss coaxial cables to the VNA ports. Thus, the transmitting horn antenna was connected to port 2, and the receiving log-periodic antenna was connected to port 1, both antennas having horizontal polarization. Then, the s12 parameter was measured, in order to assess the power of the received signal in relation to the power of the transmitted signal, in the 5–6 GHz frequency band. The distance between each of the antennas and the screen is 1.5 m, wide is enough to ensure far field propagation 1.5 m $\gg 10 * \lambda = 0.6$ m.

The first measurement setup (Fig. 2.20) was intended to assess the total attenuation (reflection and absorption) of an incident wave at the interaction with each shield. The antennas were positioned at a distance of 3 m from each other, aligning their directions of maximum radiation. Then, the shields, previously described in

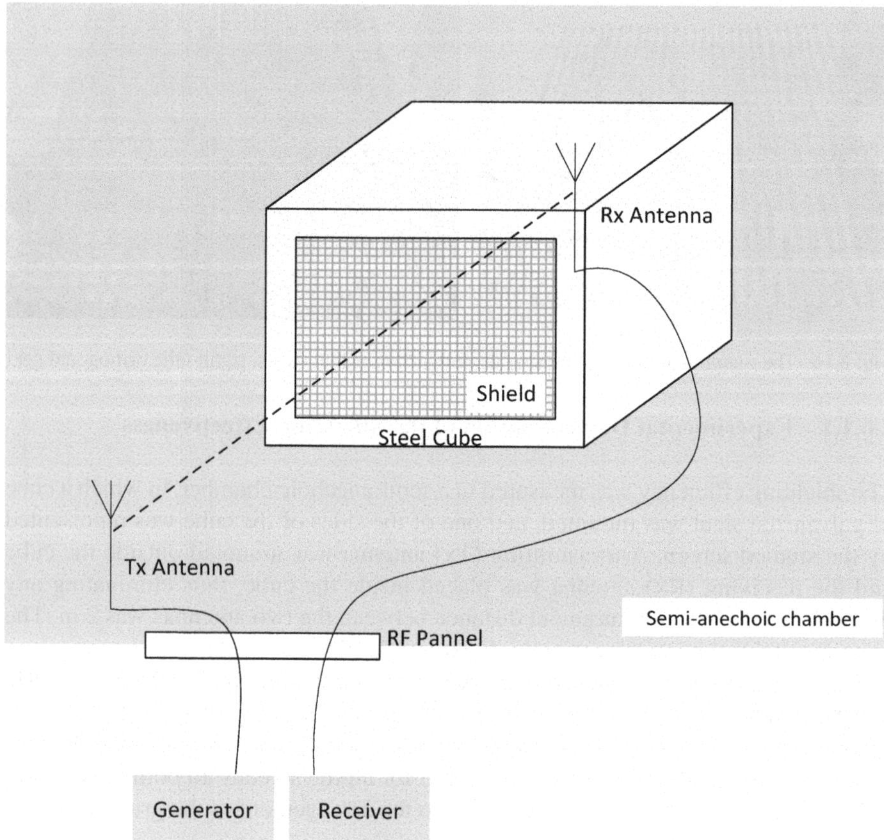

Fig. 2.17 The measurement setup for the shielding effectiveness

Sect. 2.6.1, were successively measured. The shield made of parallel strips was measured twice, by rotating it 90°, in order to obtain perpendicularity between the wave polarization and the direction of the graphite strips.

Here are the results obtained for the three cases: the graphite strips placed horizontally, then vertically, and the graphite grid were centralized and presented in Fig. 2.21. The maximum attenuation for the graphite strips placed horizontally is 7.8 dB. In this case, the graphite strips are parallel to the transmitting antenna polarization. Regarding the graphite strips placed vertically, the maximum attenuation is 2.6 dB, lower than the previous case. This is because the incident wave polarization is not aligned in the same direction as the graphite strips but is it orthogonal, and the induced motion of the charge carriers is limited. For the graphite grid, the maximum attenuation is 5.3 dB, less than expected. However, this is probably due to the fact that in the process of placing the graphite on the plexiglas in two orthogonal layers, the graphite density at the intersection of strips was lower, compared to the first screen.

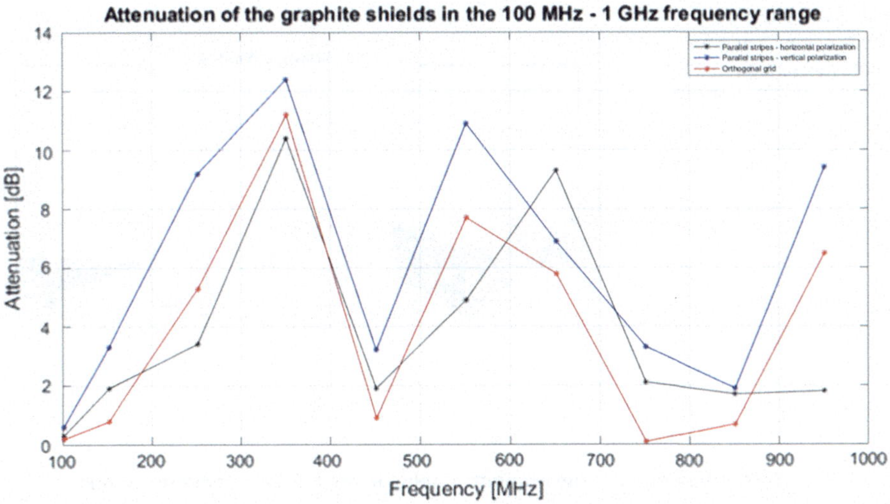

Fig. 2.18 Field attenuation in the 100 MHz–1 GHz frequency range

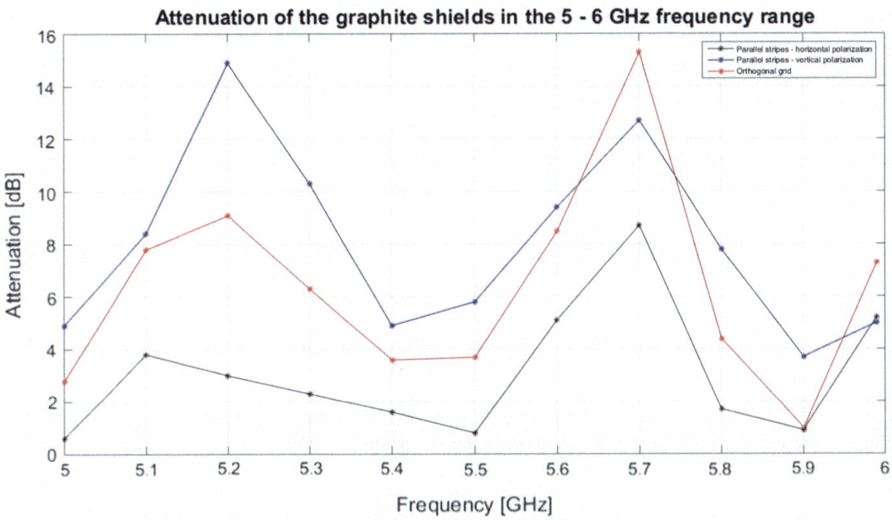

Fig. 2.19 Field attenuation in the 5–6 GHz frequency range

The second measurement setup, described in Fig. 2.22, was intended to measure only the reflected component that occurs when the incident wave encounters the shield. Both antennas are positioned in the same place, very close to each other and at 1.5 m from the shield. The measurement yields the reflection coefficient.

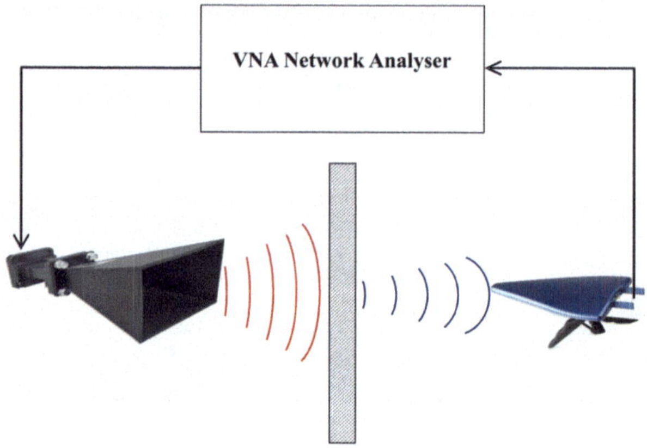

Fig. 2.20 The measurement setup for total attenuation

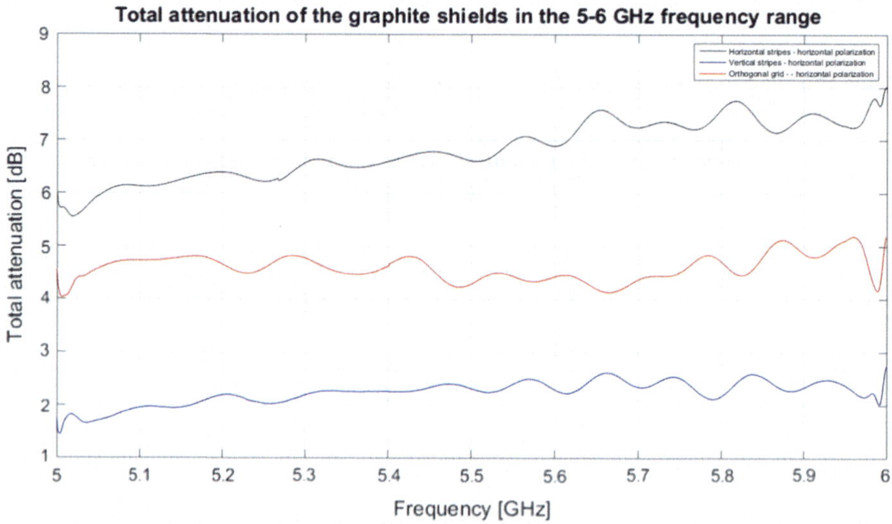

Fig. 2.21 Comparison of the total attenuation determined for the single direction graphite strips, placed horizontally (black), vertically (blue), and the graphite grid (red)

Similarly, to previous stage, in Fig. 2.23 the results of the reflection attenuation are illustrated. In this case, the maximum reflection attenuation is provided by the orthogonal grid, but the overall shielding performance is similar for all three screens in the frequency range 5–6 GHz.

From the analysis of both sets of results the relevance of the shield's orientation relative to the incident wave polarization is noted, particularly for the reflection phenomenon. Also, the screen's transparency is also relevant in the case of the total attenuation. The higher the shield's density, the better the total attenuation, meaning that most of the incident energy is absorbed into the screen's material.

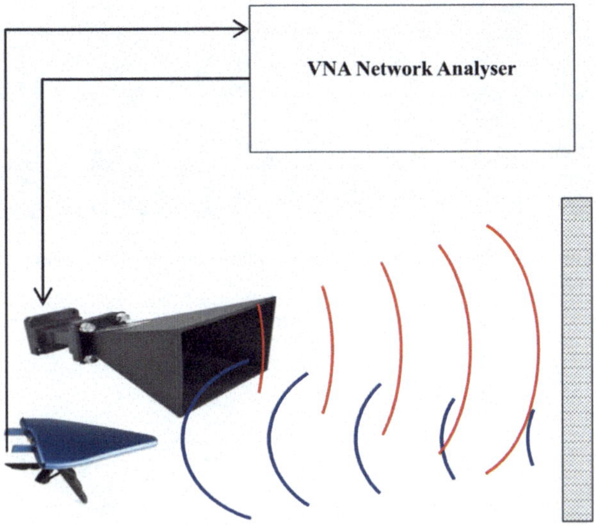

Fig. 2.22 The measurement setup for the reflection attenuation

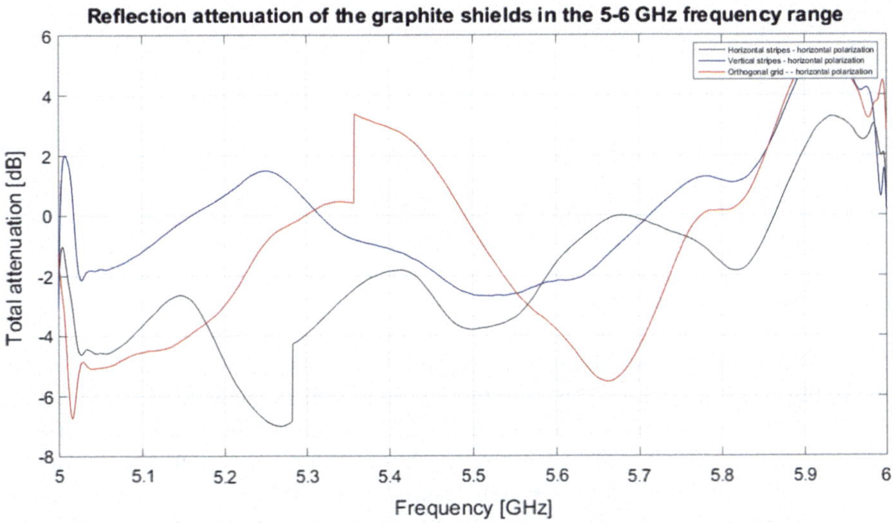

Fig. 2.23 Comparison of reflection attenuation determined for the single direction graphite strips, placed horizontally (black) and vertically (blue)

2.6.2 Experimental Analysis of the Graphite Impregnated Tweel Fabric

This paragraph investigates the use of a textile-based shielding material, with a tweel layout impregnated with graphite. Two samples of different widths of wire were used and are illustrated in Fig. 2.24.

Fig. 2.24 Photographs of the tweel fabric samples (left—sample 1, right—sample 2)

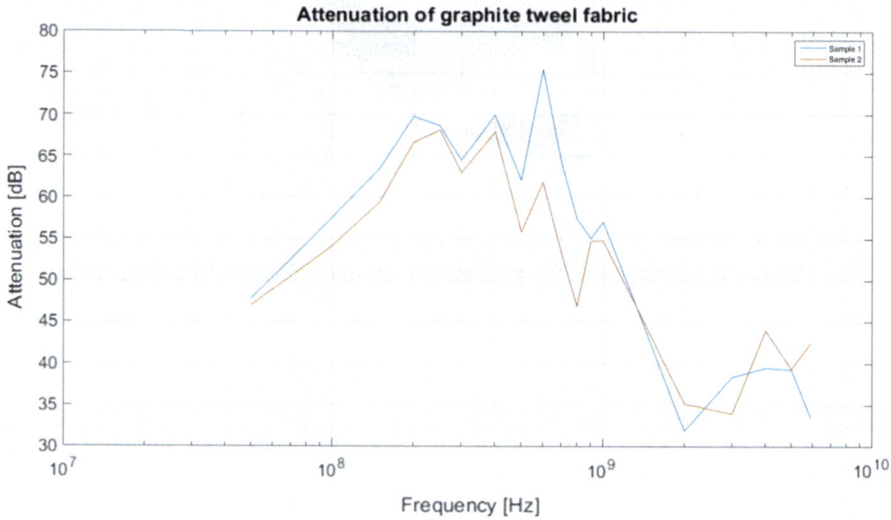

Fig. 2.25 Attenuation of the tweel fabric samples

In order to test their shielding effectiveness, a measurement setup, similar to the one in Fig. 2.17, was used. Thus, the fabric was placed as one side of a steel cube, with the receiving antenna inside the cube and the transmitting antenna outside. Both antennas were aligned at 2 m, having horizontal polarization. The results of the attenuation for both samples were measured in the 50 MHz–5.9 GHz frequency range and are illustrated in Fig. 2.25. The attenuation is very good for a textile fabric, particularly in the range of 100 MHz–1 GHz.

2.7 Potential Trends

The needs of communication and transmission of both military and civilian information has increased the frequency of communication networks—the transition to 5G technology, with trends of tens and hundreds of GHz closely followed by research on terahertz technology is already under way.

Of course, some of the materials well known for their electromagnetic properties will still be used but processed by current technologies. The first material is nanocrystalline carbon with potential applications in nanocomposite structures to ensure security by shielding in the terahertz domain but also for the realization of some components with quasi-optical applications. Experimental and theoretical research regarding the use of multilayer properties of carbon nanotube structures is sufficiently advanced in order to switch to industrial applications.

Graphene, a relatively new material discovered in the past decade, has promising properties for terahertz applications and beyond. It is expected that new properties and applications will be discovered in electronic communication technologies, electromagnetic security and other research areas. As regards to shielding, there is an interesting property of graphene foam. It is able to control the shielding factor through electric fields. Another interesting property of graphene is the fact that, unlike the materials used today, it does not reflect electromagnetic waves in the subterahertz field but absorbs them at a particularly high rate—99.99%. Its use in composite structures is promising for industrial applications related to the reduction of electromagnetic pollution. Graphene, in the form of laminated layers, together with elastic polymeric structures has properties that allow for the design of components dedicated to a specific frequency range.

References

1. Maxwell JC (1865) A dynamical theory of the electromagnetic field (PDF). Philos Trans R Soc Lond 155:459–512. https://doi.org/10.1098/rstl.1865.0008
2. Hertz HR (1887) Ueber sehr schnelle electrische Schwingungen. Ann Phys 267(7):421–448. https://doi.org/10.1002/andp.18872670707
3. Lodge OJ (1891) Experiments on the discharge of Leyden jars. Proc R Soc Lond 50:2–39. https://doi.org/10.1098/rspl.1891.0003
4. Marconi G (1897) Improvements in transmitting electrical impulses and signals, and in apparatus therefore. US patent 586193. https://patents.google.com/patent/US586193?oq=US+Patent+586193
5. Tesla N (1904) Transmission of electrical energy without wire, electrical world and engineer, March 5. http://www.tfcbooks.com/tesla/contents.htm
6. Blish JB (1899) Notes on the Marconi wireless telegraph. Proc U S Naval Inst 2(5):857–864. https://www.usni.org/magazines/proceedings/1899/october
7. Baltag O, Apreutesei AL, Rosu G et al (2019) Experimental research on textile and non-textile materials with applications to ensure electromagnetic and bio-electromagnetic compatibility. Int Conf Knowl-Based Organ 25(3):13–18. https://doi.org/10.2478/kbo-2019-0110
8. Liu L, Das A, Megaridis CM (2014) Terahertz shielding of carbon nanomaterials and their composites—a review and applications. Carbon 69:1–16. https://doi.org/10.1016/j.carbon.2013.12.021
9. Seo MA, Yim JH, Ahn YH (2008) Terahertz electromagnetic interference shielding using single-walled carbon nanotube flexible films. Appl Phys Lett 93:231–905. https://doi.org/10.1063/1.3046126
10. Polley D, Neeraj K, Barman A et al (2016) Diameter-dependent shielding effectiveness and terahertz conductivity of multiwalled carbon nanotubes. J Opt Soc Am B 33(12):2430–2436. https://doi.org/10.1364/JOSAB.33.002430

11. Zdrojek M, Bomba J, Łapińska A et al (2018) Graphene-based plastics for total sub-terahertz radiation shielding. Nanoscale 10(28):13426. https://doi.org/10.1039/C8NR02793E
12. D'Aloia AG, D'Amore M, Sarto MS (2015) Optimal terahertz shielding performances of flexible multilayer screens based on chemically doped graphene on polymer substrate. In: IEEE international symposium on electromagnetic compatibility (EMC), Dresden, Germany. https://doi.org/10.1109/ISEMC.2015.7256309
13. Tong Xu S, Fan F, Cheng J et al (2019) Active terahertz shielding and absorption based on graphene foam modulated by electric and optical field excitation. Adv Opt Mater 7(8):1–9 (1900555)
14. Zhang Y, Feng Y, Zhu B, Zhao J, Jiang T (2014) Graphene based tunable metamaterial absorber and polarization modulation in terahertz frequency. Opt Express 22(19):22743. https://doi.org/10.1364/OE.22.022743
15. Atul T, Mehrdad GN, et al (2011) Highly functionalized reactive graphene nano-sheets and films thereof. PCT/US 2011/032482, WO 2011/130507 A1
16. Kim SJ, Shin MK, Kim SH (2016) Graphene fiber and method for manufacturing same. US 2016/0318767 A1
17. Ray WJ, Lowenthal MD (2014) Graphene-based threads, fibers or yarns with nth-order layers and essensrio. US 2014/0050920 A1
18. Shah TK, Adeock DJ, Malecki HC (2011) CNT-infused fiber as a self-shielding wire for enhanced power transmission line. US 2011/0174519 A1
19. Wada T, Tsubokawa N (2012) Functional-group-modified carbon material, and method for producing same. EP 2786962 A1
20. Andryieuski A, Lavrinenko A (2013) Graphene metamaterials based tunable terahertz absorber: effective surface conductivity approach. Opt Express 21(7):9144–9155. https://doi.org/10.1364/OE.21.009144
21. Wanga L-L, Tayb B-K, Seeb K-Y, Suna Z, Tanb L-K, Luab D (2009) Electromagnetic interference shielding effectiveness of carbon-based materials prepared by screen printing. Carbon 47:1905–1910
22. Evans RW (1997) Design guidelines for shielding effectiveness, current carrying capability, and the enhancement of conductivity of composite materials. National Aeronautics and Space Administration, Marshall Space Flight Center. https://ntrs.nasa.gov/search.jsp?R=199700360552020%2D%2D01-07T12:13:02+00:00Z
23. Gaier JR (1990) Lewis Research Center, Cleveland, Ohio, NASA Technical Memorandum 103632, Intercalated graphite fiber composites as EMI shields in aerospace structures
24. Violette JLN, White DRJ, Violette MF (1987) Electromagnetic compatibility handbook. Van Nostrand, Reinhold, Co., New York
25. Rosu G, Baltag O (2019) EMI shielding disclosed through virtual and physical experiments. In: Kuruvilla J, Runcy W, Gejo G (eds) Materials for potential EMI shielding applications: processing, properties and current trends. Elsevier

Chapter 3
Carbon Fibre Reinforced Polymer Materials for Antennas and Microwave Technologies

Alexe Bojovschi, Geoffrey Knott, Andrew Viquerat, Kelvin J. Nicholson, and Tu C. Le

3.1 CFRP Material Characteristics

Carbon fibre reinforced plastic (CFRP) laminates, a well-known class of materials, are commonly used in structural and microwave applications. They are utilized for rocket motor cases, golf club, bicycles structure, airplane fuselage, terrestrial and satellite antennas. Their attractiveness is due mainly to the high stiffness to weight ratio and high strength to weight ratio, which is typically much higher than metals. Carbon fibre laminates consist of various carbon fibre plies, oriented and stacked in a prescribed pattern (analogous to plywood) tailored to meet the structural requirements. Each ply of the laminate consists of carbon fibres embedded in a plastic matrix [1, 2].

One of the primary advantages of composite structures over metallic structures is that they are more tailorable to design requirements. For example, a pressure vessel has twice the load in the hoop direction compared to the axial direction. A metallic vessel would be sized in thickness to meet the hoop load but would be oversized for the axial loads because a metal is isotropic, that is, having the same properties in all

A. Bojovschi (✉)
Center of Excellence for Technology Innovation, IntAIB, Melbourne, VIC, Australia

Department of Electronics and Communication Engineering, ASET, Amity University, Noida, India
e-mail: alexe.bojovschi@intaib.com

G. Knott · A. Viquerat
Department of Mechanical Engineering Sciences and Surrey Space Centre, University of Surrey, Guildford, UK

K. J. Nicholson
Aerospace Composite Technologies—Aerospace Division, Defence Science and Technology Group, Melbourne, VIC, Australia

T. C. Le
School of Engineering, RMIT University, Melbourne, VIC, Australia

C. Miron et al. (eds.), *Carbon-Related Materials*,
https://doi.org/10.1007/978-3-030-44230-9_3

directions [3]. However, current computational methods and advanced manufacturing techniques can be used for metallic structure design to overcome these drawbacks. On the other hand, a composite pressure vessel could be designed with just enough fibres oriented in the hoop direction to react the hoop loads and just enough fibres oriented axially to react the axial loads. Therefore, the composite pressure vessel would be a more efficient structure due to the tailorability of the composite materials. Furthermore, recent advancements in the field of nanoreinforced carbon fibres (with enhanced strength and stiffness) have further expanded the design envelope where composite structures can be applied [4]. This tailorability extends to other fields such as electromagnetics [5] where the unique anisotropic properties of carbon fibre composite structures can be employed to great effect.

The static electromagnetic properties of CFRP are in general anisotropic due to morphology of carbon fibres and the weave pattern employed in the composite structure. Furthermore, the alternating current (AC) properties of CFRP are of great interest to the microwave community. The electromagnetic properties of CFRP at microwave frequencies show a strong dependence on frequency and polarization. These properties can be employed to overcome a variety of problems in the design and deployment of microwave systems. For example, the anisotropic conductivity of unidirectional CFRP laminates can be used to minimize edge currents (on reflector dishes) or reduce cross coupling (between adjacent antenna systems). The automotive and aerospace industries have employed CFRP antennas (operating over frequencies ranging from MHz to THz) for many years that have capitalized on these anisotropic properties while also benefiting from a weight and stiffness optimized structure.

CFRP materials communally used in microwave applications consist of unidirectional fibres embedded in an epoxy matrix. The electric conductivity and permittivity of unidirectional CFRP considering the fibre volume fraction were measured over the bandwidth $10^2–10^{10}$ Hz in fibre direction and transversal to fibre direction in [6]. It was concluded that CFRP materials are good conductors up to approximately 10 GHz. Graphite, carbon-black and unidirectional CFRP are compared in [7] from DC to 10^9 Hz. Composites with interlaced carbon fibres and glass fibres are considered in [8]. The S-parameters of unidirectional and $[0\ 45\ 90\text{-}45]_{2S}$ material samples between two horn antennas were measured in [9]. In the same work, the radar cross section of CFRP parallel and perpendicular to fibre direction, both for 8.0–12.0 GHz was measured. The diamagnetic behaviour of unidirectional CFRP was measured in [10] where it was established that the real part of the permeability varies between $0 \leq \mu_r \leq 1$ depending on fibre orientation (Fig. 3.1).

3.1.1 CFRP Material Characterization with the NRW Method

One of the most effective techniques to estimate the material parameters of CFRP is to measure the S-parameters of a sample inside a waveguide. The Nicolson-Ross-Weir (NRW) equations [11, 12] can then be used to obtain conductivity, permittivity

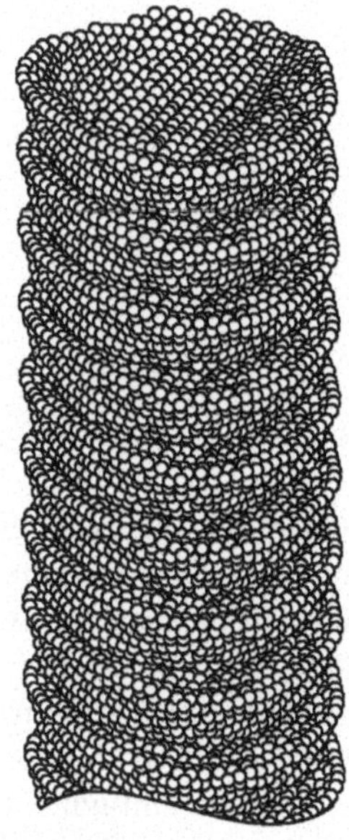

Fig. 3.1 Illustrates a carbon nanofibre as communally embedded in a polymeric matrix [4]

and permeability [6, 10] from the measured S-parameters. The NRW method was developed in the 1970s and has since become a standard method to estimate electromagnetic material properties. The first step in this method involves shifting the calibration planes to the surface of the CFRP material under test (MUT). The location of the MUT inside the test fixture may vary depending on the experimental setup and the accuracy with which the MUT can be positioned in the waveguide. This is done as indicated in [13],

$$R_i = \mathrm{e}^{-\gamma_0 L_i}, \quad i \in \{1,2\}, \tag{3.1}$$

$$S_{11}^{\mathrm{C}} = \frac{S_{11}}{R_1^2}, \tag{3.2}$$

$$S_{21}^{\mathrm{C}} = \frac{S_{21}}{R_1 R_2}, \tag{3.3}$$

where S_{11} and S_{21} are the scattering parameters from port 1 to port 1 and from port 1 to port 2 and L_i is the distance from the calibration plane at port i to the material sample. S_{11}^{C} and S_{21}^{C} are used in the NRW equation:

$$\Gamma = X \pm \sqrt{X^2 - 1}, \tag{3.4}$$

$$X = \frac{\left(S_{11}^{\mathrm{C}}\right)^2 - \left(S_{21}^{\mathrm{C}}\right)^2 + 1}{2S_{11}^{\mathrm{C}}}, \tag{3.5}$$

$$\frac{1}{\Lambda^2} = -\left(\frac{1}{2\pi d}\ln\left(\frac{1}{P}\right)\right)^2, \tag{3.6}$$

$$P = \frac{S_{11}^{\mathrm{C}} + S_{21}^{\mathrm{C}} + \Gamma}{1 - \left(S_{11}^{\mathrm{C}} + S_{21}^{\mathrm{C}}\right)\Gamma}, \tag{3.7}$$

where d is the thickness of the MUT and P is the propagation factor. The propagation factor P is a complex number in general and the logarithm in Eq. (3.6) is ambiguous by $2\pi n$, $n \in \mathrm{N}_0$. To solve this ambiguity in the NRW method, several approaches have been proposed. The most adopted ambiguity method [14] used for CFRP requires to limit the measurements to thin samples with $d < \lambda$. As CFRP are commonly built as thin laminates, this requirement can be met at microwave frequencies. Phase ambiguity can also be resolved by measuring two samples with different thickness [13]. The complex magnetic permeability μ_{r} and electric permittivity ε_{r} of CFRP laminates can be estimated according to [11, 13], as

$$\mu_{\mathrm{r}} = \mu_{\mathrm{r}}^{'} - j\mu_{\mathrm{r}}^{''} = \frac{1-\Gamma}{\Lambda(1-\Gamma)\sqrt{\frac{1}{\lambda_0^2} - \frac{1}{\lambda_{\mathrm{c}}^2}}}, \tag{3.8}$$

$$\varepsilon_{\mathrm{r}} = \varepsilon_{\mathrm{r}}^{'} - j\varepsilon_{\mathrm{r}}^{''} = \frac{\lambda_0^2}{\mu_{\mathrm{r}}}\left(\frac{1}{\lambda_{\mathrm{c}}^2} - \frac{1}{\Lambda^2}\right), \tag{3.9}$$

where λ_0 is the wavelength in free space and λ_{c} is the cut-off wavelength of the waveguide. The conductivity of CFRP can be estimated directly from S_{11} as in [15]:

$$\sigma = \frac{4\pi\mu_0 f\left(1-|S_{11}|^2\right)^2}{Z_0^2\left(\left(1+|S_{11}|^2\right) - \sqrt{-|S_{11}|^4 + 6|S_{11}|^2 - 1}\right)^2}, \tag{3.10}$$

where μ_0 is the magnetic constant, f the frequency and Z_0 is the free space impedance. Eqs. (3.8), (3.9) and (3.10) can be used for measuring complex magnetic permeability μ_{r}, electric permittivity ε_{r} *and* conductivity of unidirectional CFRP, shred-CFRP and twill-CFRP laminates.

Electromagnetic characteristics of CFRP fabrics were also measured between two loop antennas in [16]. The conductivity of CFRP laminates is estimated from the radiation efficiency of a carbon fibre monopole antenna in [17].

3.1.2 Electromagnetic Characterization of CFRP

The properties of fibres and fibre strands are in general complex quantities and by extension, so too are their macroscopic material parameters when they are embedded in a polymer matrix. Macroscopic descriptions of these composite materials use largely homogenous anisotropic properties; however, recent work on anisotropic models can provide a higher level of accuracy and can be easily implemented in computational electromagnetic modelling (CEM). A detailed discussion of equivalent layer models for CFRP is studied in [18] with different levels of detail. In [19], microscopic material properties are transformed into equivalent impedance networks. In [20], CFRP structures are considered as tensors based on initial tensorial description of CFRP electromagnetic characteristics [21]. The conductivity in the fibre direction is estimated from fibre volume fraction with a rule of mixture model. As the investigated materials are below the percolation threshold, conductivity perpendicular to fibre direction is considered very low.

CFRP materials investigated in the literature for electromagnetic applications are typically laminates with unidirectional plies (Fig. 3.3b). These plies are generally stacked with different orientations to obtain a higher structural stability, a more isotropic Young's modulus and electric conductivity. The advantage of CFRP laminates constructed from unidirectional carbon fibre plies is the simplicity of their modelling compared to those with woven or dispersed fibres. The radio frequency conductivity, permittivity and permeability of materials with short and recycled fibres significantly differ, from an electromagnetic standpoint, from unidirectional CFRP which is typically investigated and largely used in the aerospace industry.

The complex permittivity of CFRP was formulated using a permittivity tensors [21], where the direction of fibres (DF) are 45°, 0° and 90° in the plies of a laminate (Fig. 3.3b).

$$\varepsilon_{45a} = \begin{bmatrix} 0 & 0 & 0 \\ 0 & \varepsilon_{45y} & 0 \\ 0 & 0 & 0 \end{bmatrix} \quad \varepsilon_{45b} = \begin{bmatrix} \varepsilon_{45x} & 0 & 0 \\ 0 & 0 & 0 \\ 0 & 0 & \varepsilon_{45z} \end{bmatrix}, \tag{3.11}$$

$$\varepsilon_{0a} = \varepsilon_{0b} = \begin{bmatrix} \varepsilon_{0x} & 0 & 0 \\ 0 & \varepsilon_{0y} & 0 \\ 0 & 0 & 0 \end{bmatrix}, \tag{3.12}$$

$$\varepsilon_{90a} = \begin{bmatrix} 0 & 0 & 0 \\ 0 & \varepsilon_{90y} & 0 \\ 0 & 0 & \varepsilon_{90z} \end{bmatrix} \quad \varepsilon_{90b} = \begin{bmatrix} \varepsilon_{90y} & 0 & 0 \\ 0 & 0 & 0 \\ 0 & 0 & \varepsilon_{90z} \end{bmatrix}, \tag{3.13}$$

where the complex permittivities are:

$$\varepsilon_{0x} = \varepsilon_{0y} = \varepsilon_{0z} = 30 - 7.4j, \tag{3.14}$$

$$\varepsilon_{45x} = \varepsilon_{5y} = \varepsilon_{45z} = 32 - 9.2j, \tag{3.15}$$

$$\varepsilon_{90x} = \varepsilon_{90y} = \varepsilon_{90z} = 30 - 7.4j. \tag{3.16}$$

The largest contributor to the conductivity of CFRP is fibre orientation. The diagonal and off-diagonal components of the effective conductivity tensor were formulated for the CFRP [21].

$$\sigma_{45a} = \begin{bmatrix} \sigma_{45x} & 0 & 0 \\ 0 & 0 & 0 \\ 0 & 0 & \sigma_{45z} \end{bmatrix} \quad \sigma_{45b} = \begin{bmatrix} \sigma_{45x} & 0 & 0 \\ 0 & 0 & 0 \\ 0 & \sigma_{45y} & 0 \end{bmatrix}, \tag{3.17}$$

$$\sigma_{0a} = \sigma_{0b} = \begin{bmatrix} 0 & 0 & 0 \\ 0 & 0 & 0 \\ 0 & 0 & \sigma_{0z} \end{bmatrix}, \tag{3.18}$$

$$\sigma_{90a} = \begin{bmatrix} \sigma_{90x} & 0 & 0 \\ 0 & 0 & 0 \\ 0 & 0 & 0 \end{bmatrix} \quad \sigma_{90b} = \begin{bmatrix} 0 & 0 & 0 \\ 0 & \sigma_{90y} & 0 \\ 0 & 0 & 0 \end{bmatrix}, \tag{3.19}$$

where $\sigma_{45x} = \sigma_{45y} = \sigma_{45z} = \sigma_{0z} = \sigma_{90x} = \sigma_{90y} = 28,000$ S/m at 10.0 GHz. The guided wavelength in a rectangular waveguide is $\lambda_g = 1/\sqrt{\left((1/\lambda_0)^2 - (1/\lambda_c)^2\right)}$. The guided wavelength must be considered when designing a CFRP slotted waveguide antenna to ensure accurate positioning of radiative slots [22].

The tensorial formulation was shown to be the accepted method for modelling of CFRP waveguide antennas. The use of the tensorial description for carbon fibre orientation on the attenuation constant is presented in Fig. 3.2.

3.2 CFRP Materials for Automotive, Aerospace and Satellite Antennas

In automotive applications, CFRP laminates are predominantly utilized in the construction of large low curvature components (such as bonnets and roof sections). Furthermore, CFRP automotive components are typically manufactured from short and recycled carbon fibres to meet stringent repeatability and cost requirements. Hence, the tensorial electromagnetic model of CFRP might not be necessary when such automotive components are employed as the 'ground-plane' for radiating

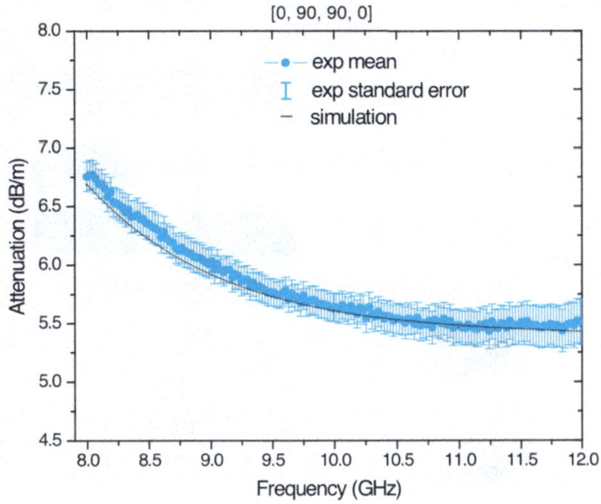

Fig. 3.2 Attenuation within CFRP waveguides with a $[0\ 90]_s$ ply stacking sequence determined experimentally and computationally

antennas. Such components may therefore be modelled as simple isotropic materials with a good approximation.

Unidirectional CFRP laminates have recently been used to fabricate microwave waveguides [21–23] where the goal was to develop lightweight, thermally stable multifunctional composite structures. A frequency selective polarizing subrefector manufactured from unidirectional CFRP was reported [24]. The work shows that the reflector can reduce the cross polarization by about 13.0 dB at S-band, and the incident angle has minimum consequence on the frequency response of the polarizer. The use of CFRP for microstrip patch, slot antenna and capacitively fed cavity-backed slot antenna has also been considered [15, 25, 26]. These studies show the advantage of backing the slots with a CFRP cavity leads to an enhancement in gain and a front-to-back ratio by 13.0 dB. The presence of defects such as cavities in dielectric materials such as epoxy resin is known to lead to partial discharge and ultimately to the breakdown of the dielectric insulator [27]. This is also applicable to CFRP laminates where the carbon fibres are encapsulated in an epoxy matrix. Methods to detect defects in CFRP have also been developed [28] to prevent imminent failures of these composites when used in critical applications such as on a fuselage of an airplane. The dimensional stability of CFRP for space antenna missions was explored [29, 30]. As the material for these applications not only should be strong, stiff and light but also have to provide dimensional stability. CFRP is shown to be one of the most promising candidates for these missions. The feasibility of using CFRP for microwave applications [21, 31, 32] for aeronautic applications (Fig. 3.3) and its structural characteristics under mechanical tests have been studied. Mathematical models and experimental validation [21] proved that CFRP can be used for waveguide structure without metallization. This helps to avoid the galvanic corrosion and minimize the costs associated with adding metallic layers.

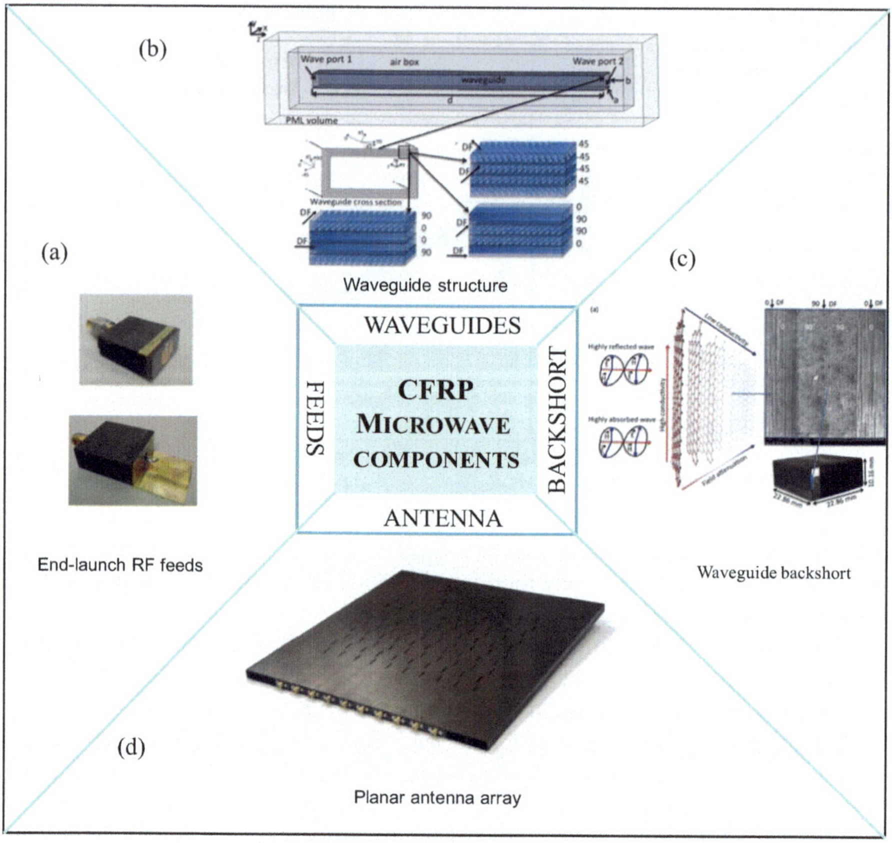

Fig. 3.3 Microwave components developed using CFRP [21, 31, 32, 40]. (**a**) Loop type end launchers Radio Frequency (RF) feeds including loop type feed and patch feed; (**b**) CFRP waveguide structure with three ply configurations $[0,90]_s$, $[90,0]_s$ and $[45,-45]_s$; (**c**) waveguide backshort with indication of the schematic figure of time harmonic waves incident on the reflective surface of the backshort and a cross section of the backshort, obtained by Scanning Electron Microscopy that indicates also the direction of fibre (DF) in the CFRP laminate and (**d**) planar antenna array with 10 × 10 waveguide slots, 10 patch feed end launchers and 10 backshorts

3.2.1 CFRP Materials for Aerospace Platforms

The integration of CFRP antennas in the structure of an aerospace systems such as the wings or the fuselage reduces the weight and the cost of airplanes. In the same time, it can provide both electromagnetic and structural functions. The aim of using CFRP materials for integrated antennas is to increase the performance of air vehicles by overcoming the limitations of traditional antennas, which increase aerodynamic drag. This concept was termed Conformal Load-bearing Antenna Structure (CLAS) [33, 34]. One implementation of CLAS for aeronautic platforms is Slotted Waveguide Antenna Stiffened Structure (SWASS) [35]. In SWASS, the top-hat stiffeners on thin skins or blade stiffeners in sandwich skins both support structural load and act as radio frequency waveguides [34, 36]. Slots cut through the outer skin and into the waveguides produce slotted waveguide antenna arrays [37]. The high specific stiffness and strength, and low susceptibility to corrosion and fatigue cracking [38–41], of CFRP material make it suitable for weight critical aeronautical application.

3.2.2 CFRP Materials for Automotive Platforms

In the coming years, the cars will change significantly. Combustion engines are already fading and a move towards electric, autonomous vehicles that move cooperatively is becoming common. The car of the future is conscious of its environment and actively protects its passengers and surroundings from damage, all while using insights form the Internet of Things (IoT) around it. Automotive antenna components are generally about 20 cm long and about 10-cm-tall casings that contain a variety of vehicular antennas and are situated on the rear of the car roof. Services delivered are mobile telephony, positioning systems, Wireless Local Area Network (WLAN), car-to-car and car-to-infrastructure communication and radio. The invention presented in [42] is related to communication antennas and particularly to antennas as employed on automobiles and trucks. The antenna disclosed is formed of fibre reinforced resin material, having reinforcing fibres in the form of strands of electrically conductive graphite, the strand being embedded under tension in the shaft of the antenna to provide high stiffness and the graphite strands being electronically connected with the metal antenna leading terminal at the base end of the antenna shaft. The performance of CFRP in a car roof compared to an aluminium sheet as antenna ground plane is assessed in [43–45]. The work indicates that CFRP with different ply structure might have a stronger influence on the radiation patterns of antennas in their vicinity. A cavity for vehicular antennas as part of a CFRP chassis that can be integrated into the car roof is presented in [46]. As automotive vehicles are increasingly equipped with CFRP components to optimize weight and maintain structural properties, CFRP antennas that have both communication and load-bearing capabilities are recently investigated [47].

3.2.3 CFRP Materials for Satellite Platforms

Future communication satellites require medium size lightweight deployable antenna reflectors which can be stowed during launch on a satellite in a minimized available envelope and are deployed during mission. The reflector according to the invention [48] is a so-called thin sheet reflector comprising sheet-like deployable panels preferably made of CFRP (Carbon Fibre Reinforced Plastic) arranged around a fixed central reflector dish as indicated in Fig. 3.4. The individual sheet-like panels

P made of CFRP in Fig. 3.4 are highly connected to a central supporting ring R which provides the structural interface to the satellite and to the tripod/feed TP assembly of the reflector as well. In addition, the central dish (CD) (Fig. 3.4) of the reflector, which is also designed as a thin-walled CFRP membrane element, is fixedly connected to the central supporting ring R. Preferably, each thin sheet-like panel P is supported by a hollow CFRP rib RB which connects a panel P to the central dish DC via mechanical bushings.

An ultra-compact deployable helical CFRP antenna, designed to enhance space-based reception of Automatic Identification System signals for maritime surveillance was investigated [29, 49, 50]. In one configuration, the helical antenna has a deployed axial length of 3.22 m, a diameter of 58 cm and a stowed (i.e. coiled) height and diameter of 5 cm, as shown in Fig. 3.5. These dimensions are optimal for the antenna to receive signals at 162 MHz. The use of ultra-thin and lighter weight CFRP composite enables an extremely high packaging ratio [51]—critical for small

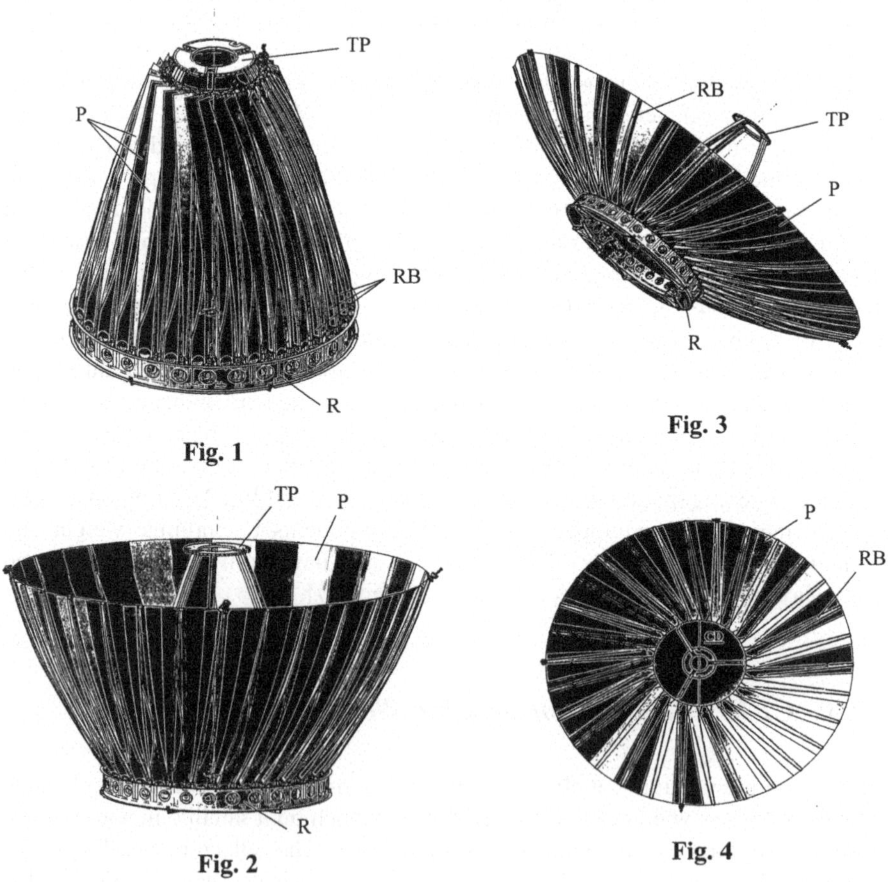

Fig. 3.4 3-D side views of the reflector in the stowed configuration (**Fig. 1**); partially deployed (**Fig. 2**); fully deployed (**Fig. 3**); top view of the reflector fully deployed (**Fig. 4**) [48]

satellite platforms and long-wavelength applications—and results in a relatively large antenna that can be stored inside a 10 cm-sided CubeSat and an estimated mass of 163 g.

3.3 CFRP Microwave Components

3.3.1 End Launchers for CFRP Antennas

The end launcher type feeds are the suitable choice for increased compactness of waveguide phase array antennas. End launchers for CFRP waveguides were developed by building on the work on various collinear end-on transitions implemented in waveguides over different frequency bands. These transitions represent the family of microwave components used in waveguide techniques. Among these, the most popular are the loop transitions [52–54] and the stepped or linearly tapered ridge waveguide sections [55, 56]. In these launchers, the grounding is attained by direct joining to the broad wall. Posts that slide through slots in the broad wall are also used for adjusting the impedance of the waveguide sections to increase matching [56]. Wheeler described the first loop transition in 1957 [57]. Later on, the loop type end-feed was modified for a multimode phase array element [58]. In 1976, Das and Sanyal devised formulas for a concentric loop transition [59].

During first works, the dimensions of the L-shaped loop were chosen so that the real part of the input impedance seen by the coaxial line was equal to the characteristic impedance of the coaxial line. The input reactance cancellation was achieved by a trial-and-error method. Deshpande together with his collaborators reported an improved formulation that takes into account the explicit expression of both real and imaginary parts of the input impedance seen by the coaxial line in terms of loop

Fig. 3.5 Deployment (unrolling) of an ultra-compact helical antenna for small satellite platforms, e.g. CubeSats [50]—a ground plane is unwrapped and then the helical boom, which is made from CFRP, provides high packaging efficiency and suitable geometry for the radiative function of the antenna element

dimensions [53]. The centric and offset loop feeds were applied in circular and rectangular waveguides [53, 54].

Traditional metallic waveguide feeds required grounding by direct electrical connection to the waveguide wall. It is not practical to do this for the CFRP waveguides because this would require the 1–3 μm resin skin surrounding each fibre to be abraded off and an electrical connection made with these fibres. This was considered unsuitable for CFRP waveguide antennas for aircraft structures [40]. Work that reported a loop type end launcher (Fig. 3.3a) optimized for a CFRP planar array panel with 9.375 GHz radiating slots was reported [32]. In this work, the requirements of keeping the integrity of the CFRP waveguide walls intact led to a loop feed that does not require grounding to the broad wall. In [60], it was also identified that a discontinuity in the broad wall reduces the structural integrity of the waveguide component.

The tensorial description of the anisotropic characteristics of CFRP [21] in the walls of the WR90 waveguide with the ply sequence $[0,90]_s$ was used to design also a patch type end launcher (Fig. 3.3a) [29]. Simulation and measurement results for the patch feed end launcher indicated the viability of CFRP patch launcher for microwave applications. The patch feed was investigated in 30 cm long $[0,90]_s$ CFRP waveguides without slots. The feed was optimized to resonate at 10 GHz. The results indicated that insertion loss is about 2.2 dB, which is due to the CFRP waveguide loss. The compactness, reproducibility and compatibility of CFRP patch launcher made it the main choice for CFRP planar antenna (Fig. 3.3d) for aeronautic applications.

3.3.2 CFRP Backshorts in Waveguide Antennas

The waveguide backshorts also named waveguide short circuits are used for radio frequency impedance matching in devices such as oscillators, detectors, mixers and multipliers, and to create standing waves in waveguides [60, 61]. Waveguide backshort are highly reflective, frequency independent, movable, located internally, lockable, not local field distorters and have good repeatability [62]. Backshorts may be in direct contact with all, some or none of the waveguide walls [62–66].

The interaction of guided waves with a contact waveguide short circuit that is constructed from CFRP (Fig. 3.3c) was reported in [31]. The behaviour of this type of backshort was simulated and measured in WR-90 CFRP waveguides. The measured average reflection coefficient and the standard error obtained when each of six CFRP backshorts (backshort-1) were placed at the reference plane [63] is illustrated in Fig. 3.6a. The CFRP and aluminium (backshort-1, 2 and 4) backshorts show good characteristics with the magnitude of the reflection coefficient between 0.97 and 1. The metallic backshorts have higher reflection coefficient by about 0.007 compared to CFRP (Fig. 3.6a).

The slightly smaller values of the reflection coefficient are mainly due to losses in the skin depth of CFRP. The losses are the time-average Joule's losses for the conductors and time-average power loss for the dielectric. The reflected phase angle varies from about 179.00 at 8 GHz to about 178.25 at 12 GHz. The deviation of about 0.75 over the X-band may be ignored [63, 64] because the accuracy of the Thru-Reflect-Line (TRL) calibration is 0.05 dB in magnitude and 1 m phase over the measurement bandwidth. In this case, the phase over the X-band for these waveguide backshorts can be regarded, in agreement with the literature [65, 66], as 'constant phase'.

The performance of the CFRP backshorts is shown to be comparable, and in some cases, exceed that of the conventional backshorts [31]. The measured reflection coefficient for CFRP backshorts is approximately 0.98 at the reference plane and transmission less than 0.001 over the X-band. The advantages of the CFRP backshort in CFRP structures are compatibility, lighter weight than a metallic counterpart and resistant to galvanic deterioration. The importance of the work was for the first time proposing a suitable short circuit that can be used for lightweight microwave applications such as CFRP-based waveguide array antennas.

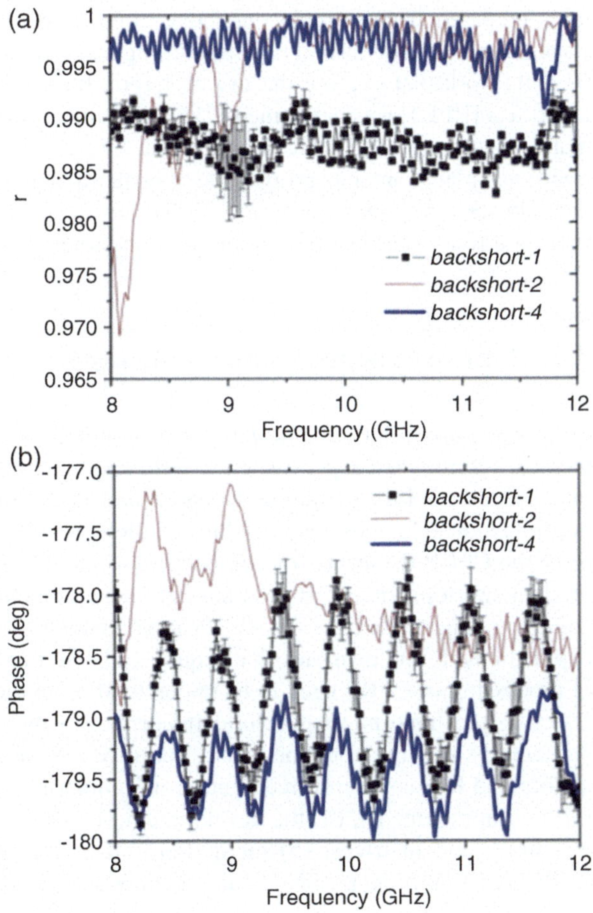

Fig. 3.6 The measured reflection coefficient (**a**) and phase of the reflected wave (**b**) at the reference plane for six CFRP backshorts (backshort-1) and conventional aluminium backshorts (backshort-2 and -4) [31]

3.3.3 CFRP Materials for Antennas

Antennas are a cornerstone of modern wireless communication technology. They are designed to receive and emit electromagnetic radiation and to act as a medium between free space and wireless devices. A minimum requirement of conventional antennas is that they comprise an electrical conductor. For this reason, most conventional antennas have been limited to metallic structures. For antenna applications in which weight optimization is required, metallic antennas can be problematic. In diverse structural applications, polymers and polymer composites have been used as a lightweight substitute for metals. Certain polymers and polymer composites are electrically conductive or can be made electrically conductive; however, low conductivities have generally limited their use as a metal replacement in applications requiring electrical conductivity. Non-metallic or at least partially non-metallic antenna structures would be of considerable use in a variety of applications in which metallic antennas are conventionally used. To provide alternative approach to traditional antennas that are wholly metallic solutions were proposed for antenna structures prepared from highly conductive polymer composites utilizing conductive carbon nanotubes or carbon fibre as a filling material in a polymeric matrix [21, 42, 67–70]. Work that explore the use of carbon fibre composites in radio frequency identifiers [RFID] was performed [71]. The work reported in [40, 41] demonstrated that a structural panel manufactured from CFRP (Fig. 3.3d) can also behave as a traditional planar antenna array made of metal such as aluminium. The work opened the path for developing conformal CFRP antennas with both electromagnetic and structural role in aeronautics, automotive and satellite systems.

3.4 Future Prospects and Challenges

Innovative carbon-based materials can be competitive to conventional materials in the near future when aspects such as lightweight, strength and life cycle are taken into account. With continuous research and development, it is anticipated that multiple industries will benefit. Many challenges still need to be overcome to optimize the CFRP materials for engineering usage. Creativity, design, research and product development efforts and business development skills are required to overcome existing challenges in CFRP-based industries and to develop innovative applications. Some of issues that require to be addressed include: (1) the science of processing of CFRP need to be understood more thoroughly, especially factors affecting the nano-microstructural integrity; (2) the damage tolerant properties particularly fracture toughness and ductility that need improvement; (3) high-quality and low-cost reinforcements from industrial wastes and by-products need further improvements; (4) the development of hybrids based on non-standard carbon fibres and matrices; (5) methods to effectively classify different grades of CFRP based on property profile and manufacturing cost; and (6) the development of simple, economical and portable non-destructive kits to quantify undesirable defects in carbon fibre-based composites. CFRP was proven to be effective for designing automotive components to reduce weight and to increase passenger's safety. For example, the specific energy absorption of carbon fibre reinforced thermoplastic composites in car crash tests [72] was shown to be much higher than that of aluminium alloys. Future work has the potential to be of benefit to transport industries where accidents are common.

Developments not only in military industry but also for commercial passenger airplanes initiated by Airbus [73] will continue to improve aspects such as safety and environmental impact. The improvement in material properties has helped us to travel quickly and economically around the world, by improving the performance and operations of modern aircrafts. Airbus extends its innovation in composite materials through their usage in applications such as flying drones for commercial use in the near future.

One of the future usages of CFRP is for adaptive structures that can purposely vary their geometric configuration and physical properties [74] for space applications. Beyond the current applications for the (Automatic Identification System) AIS satellite, such as the development of the structural design of deployable helical antennas made from fibre composite materials [75], and recent use of CFRP [50] provides a promising concept for future satellites.

The future challenges revolve not only around developing more advanced CFRP-based technologies but also considering their environmental effects. For example, CFRP-based drone systems were indicated to not only disrupt bird's habitat and life but also to cause injuries such as those of eagles [76]. System thinking methods need to be employed also to help optimize the impact of emerging CFRP-based technology on the environment. The design and synthesis of materials with useful, novel properties is one of the most active areas of contemporary science, generating a veritable explosion of scientific activity [77]. The ability to predict the properties of novel materials prior to manufacture and to understand the relationships between the microscopic properties of molecular components and the macroscopic materials properties, provided by Machine Learning (ML) is of substantial benefit to materials designers. The use of these ML and Artificial Intelligence (AI) to design smart CFRP materials with mechanical, electromagnetic and environmental characteristics will play a key role to solve current challenges and redefine future applications.

References

1. Legein F, Lippens P, Pauwels M, Vanlandeghem A (2009) Method for pre-treating fibre reinforced composite plastic materials prior to painting and method for applying a painting layer on fibre reinforced composite plastic materials. Patent no. US 2009/0311442 A1
2. Nobuyuki A, Norimitsu N, Kenichi Y, Junko K, Hiroshi T (2013) A Prepreg and carbon fibre reinforced composite materials. Patent no. EP 2,666,807 A2
3. Rese RP, Gossard TW Jr (1999) Near zero CTE carbon fibre hybrid laminate. Patent no. US 5,993,934

4. Kruckenberg TM, Hill VA (2011) Method of making nanoreinforced carbon fibre and components comprising nanoreinforced carbon fibres. Patent no. US 2011/0001086 A1, 2011
5. Zhao Q, Zhang K, Zhu S, Xu H, Cao D, Zhao L, Zhang R, Yin W (2019) Review on the electrical resistance/conductivity of carbon fibre reinforced polymer. Appl Sci 9:1–2
6. Kim HC, See SK (1990) Electrical properties of unidirectional carbon-epoxy composites in wide frequency band. J Phys D Appl Phys 23(7):916
7. Ezquerra TA, Connor MT, Roy S, Kulescza M, Fernandes-Nascimento J, Baltá-Calleja FJ (2001) Alternating-current electrical properties of graphite, carbon-black and carbon-fibre polymeric composites. Compos Sci Technol 61(6):903–909
8. Lee SE, Oh KS, Kim CG (2006) Electromagnetic characteristics of frequency selective fabric composites. Electron Lett 42(8):439–441
9. Galehdar A, Nicholson KJ, Rowe WST, Ghorbani K (2010), The conductivity of unidirectional and quasi isotropic carbon fibre composites. In: The 40th European microwave conference, pp 882–885
10. Galehdar A, Nicholson KJ, Callus PJ, Rowe WST, John S, Wang CH, Ghorbani K (2012) The strong diamagnetic behaviour of uni-directional carbon fibre reinforced polymer laminates. J Appl Phys 112(11):113921
11. Nicolson AM, Ross GF (1970) Measurement of the intrinsic properties of materials by time-domain techniques. IEEE Trans Instrum Meas 19(4):377–382
12. Weir WB (1974) Automatic measurement of complex dielectric constant and permeability at microwave frequencies. Proc IEEE 62(1):33–36
13. Vicente AN, Dip GM, Junqueira C (2011) The step by step development of NRW method. In: 2011 SBMO/IEEE MTT-S international microwave and optoelectronics conference (IMOC 2011), pp 738–742. https://doi.org/10.1109/IMOC.2011.6169318
14. Ghodgaonkar DK, Varadan VV, Varadan VK (1990) Free-space measurement of complex permittivity and complex permeability of magnetic materials at microwave frequencies. IEEE Trans Instrum Meas 39(2):387–394
15. Galehdar A, Rowe WST, Ghorbani K, Callus PJ, John S, Wang CH (2011) The effect of ply orientation on the performance of antennas in or on carbon fibre composites. Prog Electromagn Res 116:123–136
16. Doan T, Walters A, Leat C. (2009) Characterisation of electromagnetic properties of carbon fibre composite materials. In: Electromagnetic compatibility symposium Adelaide, pp 87–91
17. Galehdar A, Callus PJ, Ghorbani K (2011) A novel method of conductivity measurements for carbon-fibre monopole antenna. IEEE Trans Antennas Propag 59(6):2120–2126
18. Holloway CL, Sarto MS, Johansson M (2005) Analyzing carbon-fibre composite materials with equivalent-layer models. IEEE Trans Electromagn Compat 47(4):833–844
19. Wasselynck G, Trichet D, Ramdane B, Fouldagar J (2010) Interaction between electromagnetic field and CFRP materials: a new multiscale homogenization approach. IEEE Trans Magn 46(8):3277–3280
20. Fouladgar J, Wasselynck G, Trichet D (2013) Shielding and reflecting effectiveness of carbon fibre reinforced polymer (CFRP) composites. In: International symposium on electromagnetic theory, pp 104–107
21. Bojovschi A, Nicholson KJ, Galehdar A, Callus PJ, Ghorbani K (2012) The role of fibre orientation on the electromagnetic performance of waveguides manufactured from carbon fibre reinforced plastic. PIER B 39:267–280
22. Bojovschi A, Rudd M, Scott J, Ghorbani K (2017) The guide wavelength in a CFRP WR90 waveguide. In: Proceedings of Asia Pacific microwave conference
23. Gray D, Nicholson KJ, Ghorbani K, Callus PJ (2010) Carbon fibre reinforced plastic slotted waveguide antenna. In: Proceedings of Asia Pacific microwave conference, pp 307–310
24. Galehdar A, Rowe WST, Ghorbani K, Callus PJ, John S, Wang CH (2012) A frequency selective polarizer using carbon fibre reinforced polymer composite. Prog Electromagn Res C 25:107–118

25. Mehdipour A, Sebak A-R, Trueman CW, Rosca ID, Hoa SV (2011) Performance of microstrip patch antenna on a reinforced carbon fibre composite ground plane. Microw Opt Technol Lett 53(6):1328–1331
26. Galehdar A, Callus PJ, Rowe WST, Wang CH, John S, Ghorbani K (2012) Capacitively fed cavity-backed slot antenna in carbon-fibre composite panels. IEEE Antenna Wireless Propag Lett 11:1028–1031
27. Bojovschi A, Rowe WR, Wong KL (2010) Electromagnetic field intensity generated by partial discharge in high voltage insulating materials. Prog Electromagn Res 104:167–182
28. Megali G, Pellicano D, Cacciola M, Calcagno S, Versaci M, Morabito FC (2010) EC modeling and enhancement signals in CFRP inspection. Prog Electromagn Res M 14:45–60
29. Knott G, Viquerat Andrew A, Bojovschi A (2018) Design of deployable helical antennas for space-based automatic identification system reception. In: Proceedings of emerging sensing technologies summit 2018, pp 1–9
30. Sanjuan J, Preston A, Korytov D, Spector A, Freise A, Dixon G, Livas J, Mueller G (2011) Carbon fibre reinforced polymer dimensional stability investigations for use on the laser interferometer space antenna mission telescope. Rev Sci Instrum 82:124501-1–124501-11
31. Bojovschi A, Scott J, Ghorbani K (2013) The reflectivity of carbon fibre reinforced polymer short circuit illuminated by guided microwaves. Appl Phys Lett 103:111910-1–111910-5
32. Bojovschi A, Gray D, Ghorbani K (2013) A loop-type end launcher for carbon fibre reinforced polymer waveguides. PIER M 31:13–27
33. Lockyer AJ, Alt KH, Coughlin DP, Durham MD, Kudva JN, Goetz AC, Tuss J (1999) Design and development of a conformal load-bearing smart skin antenna: overview of the AFRL smart skin structures technology demonstration (S^3TD). Proc SPIE 3674:4010–4024
34. Callus PJ (2008) Novel concepts for conformal load-bearing antenna structure. Report No. DSTO-TR-2096, Defence Science and Technology Organisation, Australia
35. Callus PJ, de LaHarpe JCD, Tuss JM, Baron WG, Kuhl DG (2012) Slotted waveguide antenna stiffened structure. United States Patent No. 8149177, April 3
36. Nicholson KJ, Callus PJ (2010) Antenna patterns from single slots in carbon fibre reinforced plastic waveguides. Report No. DSTO-TR-2389, Defence Science and Technology Organisation, Australia
37. Stevenson AF (1948) Theory of slots in rectangular waveguides. J Appl Phys 19:24–38
38. Golfman Y (2011) Hybrid anisotropic materials for structural aviation parts. Taylor & Francis Group, Boca Raton
39. Niu MCY (1996) Composite airframe structures, 2nd edn. Conmilit Press Ltd., Hong Kong
40. Callus PJ, Nicholson KJ, Bojovschi A, Ghorbani K, Baron W, Tuss J (2012) A planar antenna array manufactured from carbon fibre reinforced plastic. In: 28th International congress of the aeronautical sciences, pp 1–10
41. Bojovschi A, Nicholson KJ, Ghorbani K (2018) Load bearing slotted waveguide carbon fibre reinforced polymer antenna stiffened structure. In: Emerging sensing technologies summit 2018
42. DeLoach TR, Kusek WW (1979) Whip antenna formed of electronically conductive graphite strands embedded in a resin material. Patent no. US 4,134,120
43. Artner G, Langwieser R (2017) Performance of an automotive antenna module on a carbon-fibre composite car roof. In: 10th European conference on antennas and propagation (EuCAP), pp 1–4
44. Ekiz L, Thiel A, Klemp O, Mecklenbrauker CF (2013) MIMO performance evaluation of automotive qualified LTE antennas. In: 7th European conference on antennas and propagation (EuCAP)
45. Artner G, Langwieser R, Lasser G, Mecklenbrauker CF (2014) Effect of carbon-fibre composites as ground plane material on antenna performance. In: IEEE-APS topical conference on antennas and propagation in wireless communications (APWC)
46. Artner G, Langwieser R, Mecklenbräuker CF (2017) Concealed CFRP vehicle chassis antenna cavity. IEEE Antennas Wireless Propag Lett 16:1415–1418

47. Artner G, Kotterman W, Galdo GD, Hein MA (2018) Conformal automotive roof-top antenna cavity with increased coverage to vulnerable road users. IEEE Antennas Wireless Propag Lett 17(12):2399–2403
48. Schmid M, Scheulen D, Barho R, Weimer P (2004) Deployable antenna reflector. Patent no. EP 1 386 838 A1
49. Knott G, Wu C, Viquerat Andrew A (2019) Deployable bistable composite helical antennas for small satellite applications. In: Proceedings of AIAA SciTech 2019 Forum, AIAA. https://doi.org/10.2514/6.2019-1260
50. Knott G, Viquerat A (2019) An ultra-compact helical antenna for small satellites. In: 70th international astronautical congress. International Astronautical Federation, Washington, DC
51. Knott G, Viquerat A (2019) Helical bistable composite slit tubes. Compos Struct 207:711–726
52. Chan KK, Martin R, Chadwick K (1998) A broadband end launcher coaxial-to-waveguide transition for waveguide phased arrays. In: Proceedings of IEEE, pp 1390–1393
53. Deshpande MD, Das BN, Sanyal GS (Aug. 1979) Analysis of an end launcher for an X-band rectangular waveguide. IEEE Trans Microw Theory Tech 27(8):731–735
54. Saad SM (Feb. 1990) A more accurate analysis and design of coaxial-to-rectangular waveguide end launcher. IEEE Trans Microw Theory Tech 38(2):129–134
55. Levy R, Hendrick LW (2002) Analysis and synthesis of in-line coaxial-to-waveguide adapters. In: Proceedings of IEEE microwave symposium, Seattle, USA, pp 809–811
56. Dix JC (1963) Design of waveguide/coaxial transition for the band 2.5–4.1 Gc/s. Proc Inst Electr Eng 110(2):253–255
57. Wheeler GJ (1957) Broad band waveguide to coaxial transitions. IRE Convention Record Part 1, pp 182–185
58. Tang R, Wong NS (1968) Multimode phased array element for wide scan angle impedance matching. Proc IEEE 56:1951–1959
59. Das BN, Sanyal GS (1963) Coaxial to waveguide transition (end launcher type). In: Proceedings of the Institute of Electrical Engineers, London, vol 110, pp 253–255
60. Ellison BN, Little LT, Mann CM, Matheson DN (1991) Quality and performance of tunable waveguide backshorts. Electron Lett 27:139–141
61. Bilik V, Bezek J (2006) Noncontacting R26-waveguide sliding short for industrial applications. In 2006 European microwave conference, Manchester, pp 1032–1035
62. Eisenhart RL, Monzello RC (1982) A better waveguide short circuit. In: IEEE MTT-S international microwave symposium digest, Dallas, TX, USA, pp 360–362
63. Brewer MK, Raisanen AV (1982) Dual-harmonic noncontacting millimeter waveguide backshorts: theory, design, and test. IEEE Trans Microwave Theory Tech 30(5):708–714
64. Weller TM, Katehi LPB, McGrath WR (1995) Analysis and design of a novel noncontacting waveguide backshort. IEEE Trans Microw Theory Tech 43(5):1023–1030
65. McGrath WR, Weller TM, Katehi LPB (1995) A novel noncontacting waveguide backshort for submillimeter wave frequencies. Int J Infrared Millim Waves 16:237–256
66. Kerr AR (1988) An adjustable short-circuit for millimeter waveguides. Electronics Division Internal Report No. 280. National Radio Astronomy Observatory, Charlottesville, VA
67. Gould WI Jr., Evans J (1973) Millimeter wave antenna system. Patent no 3,716,869
68. Jonda W (1978) Lightweight structural part formed of carbon fibre-reinforced plastic. Patent no US 4,092,453
69. Curran S, Talla J, Dias S (2012) Antennas based on a conductive polymer composite and method for production thereof. Patent no. US 8,248,305 B2
70. Konanur AS, Karacaoglu U (2016) Magnetic field pass through surfaces in carbon fibre reinforced polymer. Patent no. US 9,252,482, B2
71. Mehdipour A, Rosca ID, Sebak A-R, Trueman CW, Hoa SV (2010) Advanced carbon-fibre composite materials for RFID tag antenna applications. ACES J 25(3):218–229
72. Jerome P (2001) Composite materials in the Airbus A380-from history to future. In: Proceedings of 13th ICCM conferences, Beijing, China

73. Miura K, Furuya H (1988) Adaptive structure concept for future space applications. AIAA J 26(8):994–1002
74. Sproewitz T, Block J, Bäger A, Hauer L, Schuetze M (2011) Deployment verification of large CFRP helical high-gain antenna for AIS signals. In: Proceedings of aerospace conference, pp 1–12
75. Rebolo-Ifrán N, Grilli MG, Lambertucci SA (2019) Drones as a threat to wildlife: YouTube complements science in providing evidence about their effect. Environ Conserv 46(3):205–210
76. Le T, Epa VC, Burden FR, Winkler DA (2012) Quantitative structure–property relationship modeling of diverse materials properties. Chem Rev 112(5):2889–2919
77. Le TC, Winkler DA (2016) Discovery and optimization of materials using evolutionary approaches. Am Chem Soc 116(10):6107–6132

Chapter 4
Structural Design and Optimization of Slotted Waveguide Antenna Stiffened Structures under Compressive Load

Woon Kim, Robert A. Canfield, William Baron, James Tuss, and Jason Miller

4.1 Introduction

Slotted waveguide antenna stiffened structures (SWASS) have been developed with the aim of improving the structural strength and stiffness of an integrated aircraft wing or fuselage, while evaluating electromagnetic (EM) or radio frequency (RF) performance. The essence of SWASS is to create a Conformal Load-Bearing Antenna Structure (CLAS) by turning the skins of an aircraft structure into a radar array.

Figure 4.1 illustrates an example of an antenna-integrated structure. A major benefit is the weight savings. Conventional radar installed into an aircraft radome minimize aerodynamic loading effects; however, it is rather bulky and adds significant weight to the aircraft. Due to its honeycomb structure composition, the SWASS has the potential to lower the overall aircraft structural weight, while maintaining structural strength. Furthermore, SWASS utilizes the whole aircraft skin structure as a radar antenna which can significantly enhance the radar performance. SWASS may be categorized as an advanced version of CLAS [1], which has received much attention due to its favorable characteristics, such as multifunctional integrity of

The views and conclusions contained herein are those of the authors and should not be interpreted as necessarily representing the official policies or endorsements, either expressed or implied, of Air Force Research Laboratory or the US Government. Cleared for Public Release 88ABW-2013-2532.

W. Kim · R. A. Canfield (✉)
Department of Aerospace and Ocean Engineering, Virginia Tech, Blacksburg, VA, USA
e-mail: bob.canfield@vt.edu

W. Baron · J. Tuss
Air Force Research Laboratory, WPAFB, OH, USA

J. Miller
Booz Allen Hamilton, Dayton, OH, USA

C. Miron et al. (eds.), *Carbon-Related Materials*,
https://doi.org/10.1007/978-3-030-44230-9_4

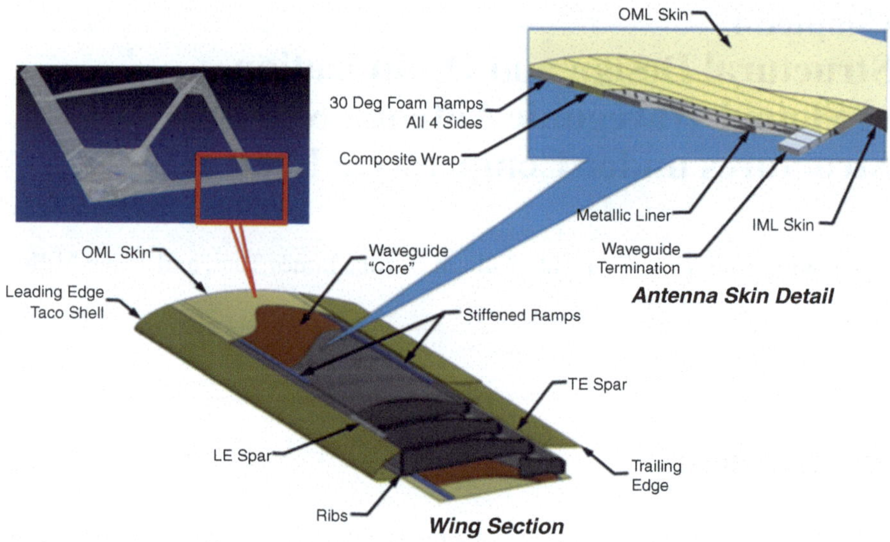

Fig. 4.1 SWASS integration scheme

aircraft and antenna structures. Callus [2] presented a brief history of the CLAS concept and reviewed various types of antenna-integrated structures.

The continuing work on SWASS has focused on improving structural performance by enhancing robustness and stability. Sabat and Palazotto [3] studied structural failure and potential structural instability of SWASS through a single composite rectangular wave guide. Simulation results under uniaxial compressive load were verified against experimental results. Kim et al. [4] presented modeling and analysis of four novel SWASS design concepts.

Ha and Canfield [5] and Kim et al. [6] suggested an optimum design technique for WR-90 waveguide to maximize its EM performance. The authors investigated the trends of radio wave pattern sensitivity by varying the geometrical parameters of the waveguide. Regarding RF performance sensitivity associated with the aircraft structure, Smallwood et al. studied the dipole antenna-embedded structure on joined-wing aircraft [7]. The authors investigated the effect on sensitivity of RF performance due to structural deformation associated with aerodynamic loading. Knutsson et al. sought the materials not only to save weight, but also to increase electrical properties [8].

The SWASS design was formulated to minimize weight and maximize structural strength. This chapter presents three major findings: (1) low-fidelity 2D and high fidelity 3D composite modeling and verification, (2) configuration of four design concepts of SWASS, and (3) nonlinear analysis for optimal design and experimental validation of the four SWASS concepts.

4.2 Design Concepts of SWASS

Structural analysis and evaluation of four novel design concepts for SWASS, shown in Fig. 4.2, was carried out. These various composite structures possess a spectrum of performance capabilities characterized by (1) structural stability subject to compressive loading conditions, (2) effective mass ratio compared to the metallic reinforcement, and (3) conductivities to transmit an EM signal.

In more detail, Concept 1 has load-bearing carbon fiber waveguide tubes enclosed by carbon fiber inner and outer mold line face sheets. This structure is expected to effectively support various loads and prevent buckling and fiber strain failure. In addition, due to conductivity of carbon fiber, EM signals can be guided by the fiber waveguide tube. Because the carbon fiber face sheets are also electrically conducting, slots must be cut through the top skins, which weaken structural stiffness. Concept 2 is similar to Concept 1, except that its face sheet material is fiberglass. Fiberglass is electromagnetically transparent; therefore, no additional slots are required on the top skin, thereby increasing structural strength, even though fiberglass is not considered as stiff as carbon fiber.

Concept 3 is different from Concepts 1 and 2 in that it has thin copper metallic foil waveguide tubes to conduct EM signals. Only the metallic foil waveguide has slots, which has relatively less influence on mechanical strength. Face sheets are fiberglass, which is electromagnetically transparent; therefore, no additional slots are required through the outer skins. Concept 4 is same as Concept 3, except for

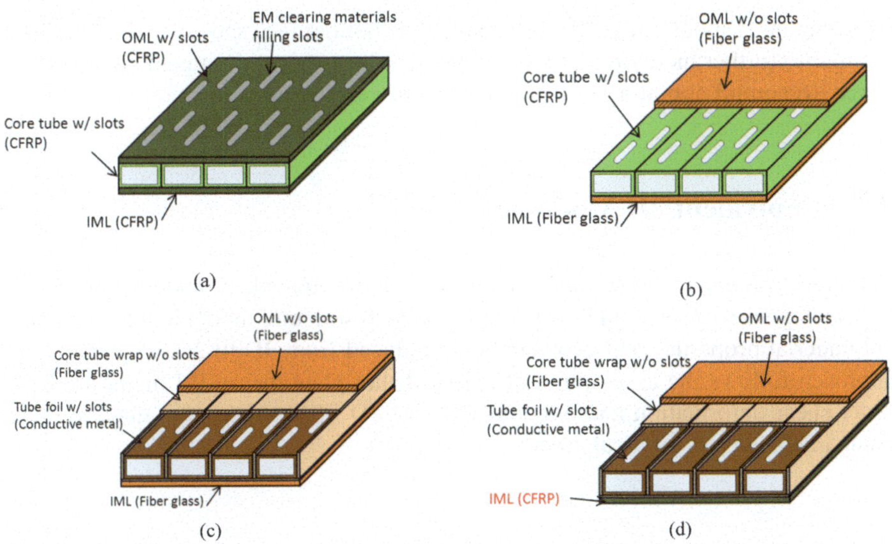

Fig. 4.2 Four design concepts of slotted waveguide antenna stiffened structures. (**a**) Concept 1: carbon tube with carbon skins. (**b**) Concept 2: carbon tube with fiber glass skins. (**c**) Concept 3: metallic foil WG: Foil wrapped by fiber glass with fiber glass skins. (**d**) Concept 4: metallic foil WG: Foil wrapped by fiber glass with fiber glass (OML) and CFRP (IML) skins

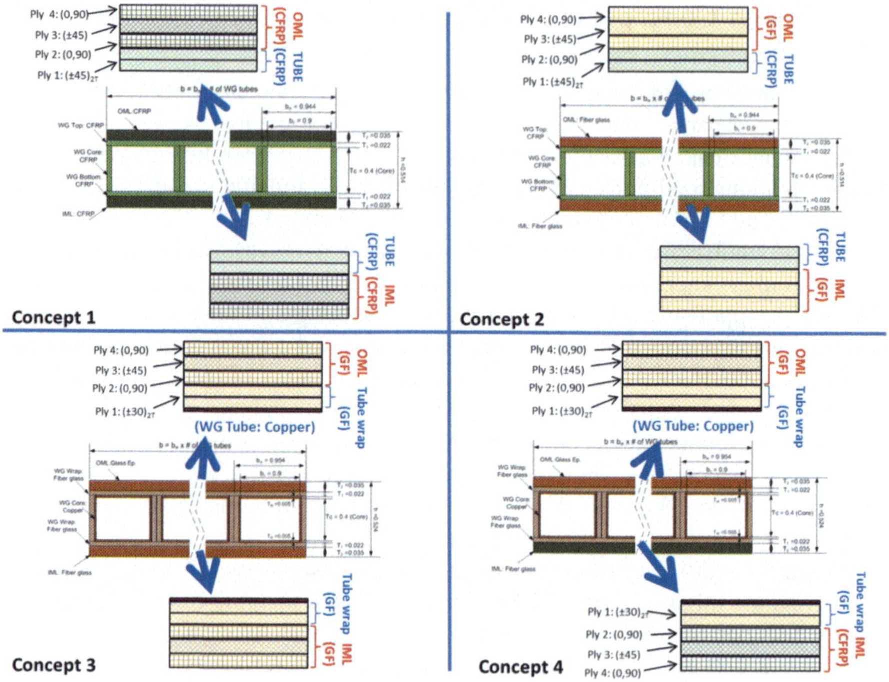

Fig. 4.3 Lay-up configuration of four SWASS concepts

replacing inner mold line (IML) material on the bottom face sheet with carbon fiber in order to further increase stiffness. Figure 4.3 illustrates the detailed lay up of Air Force Research Laboratory (AFRL) SWASS test articles.

4.3 Equivalent 2D Modeling

In this section, the mathematical derivation for a 2D equivalent model is presented. It is analogous to a smeared honeycomb sandwich structure model in that the inherent material properties and geometries are reduced from 3D to 2D. The analytical approach follows the assumption of classical thin plate theory [9] including first-order shear deformation and is compared to a 2D FEM model that may be more comparable to the analytical model.

4.3.1 Equivalent Modeling of Core Webs

For model comparison and verification, two configurations of equivalent model were examined, which simplified the stiffness from the waveguide tubes: firstly, equivalent core in Fig. 4.4a and secondly, equivalent beam stiffness in Fig. 4.4b. The

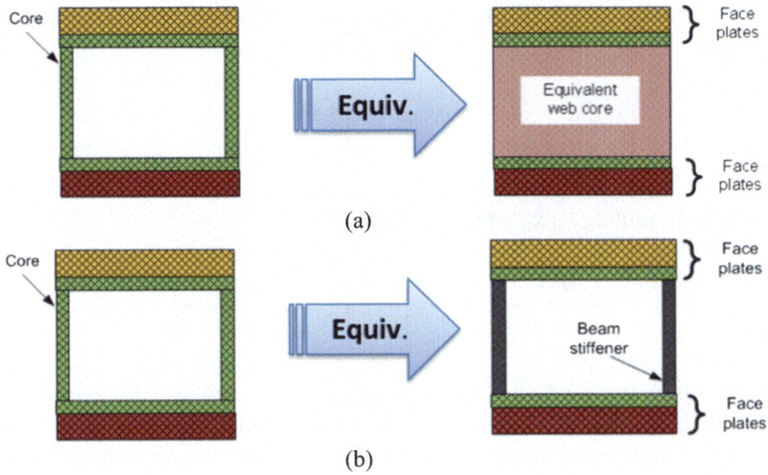

Fig. 4.4 Two configurations of equivalent model. (**a**) Equivalent core. (**b**) Equivalent beam stiffness

first configuration smears the waveguide web material across a homogenous core, while the second configuration models the waveguide core as beam stiffeners.

The benefit of these approaches is to simplify the complex 3D composite structure for SWASS, such as composite web columns in internal structure or slots on the surface plate into the corresponding 2D plate or 1D beam material problems. Consequently, another advantage is to save computational cost. These simplified configurations were considered as a possible preliminary step to smearing the effect of the slots as a further simplification. Theoretical results demonstrate the accuracy and establish a guideline for suitability of an equivalent model.

4.3.2 Mathematical Formation

The general expression for 2D equivalent plate model is derived from the principal virtual work problem [10–12].

$$\delta U + \delta V = \delta \Pi = 0, \tag{4.1}$$

where $U + V = \Pi$ is the total potential energy, δU the virtual strain energy, δV the virtual work done applied forces. Specifically, the virtual strain energy (δU) is written as

$$\begin{aligned}\delta U = \int_V [\sigma_{xx}\delta\varepsilon_{xx} + \sigma_{yy}\delta\varepsilon_{yy} \\ + 2\sigma_{yz}\delta\gamma_{yz} + 2\sigma_{xz}\delta\gamma_{xz} + 2\sigma_{xy}\delta\gamma_{xy}]dV.\end{aligned} \tag{4.2}$$

Classical laminate plate theory dictates that the strain components can be expressed as

$$\varepsilon_{xx} = \varepsilon_{xx}^{0} + z\kappa_{x}, \varepsilon_{yy} = \varepsilon_{yy}^{0} + z\kappa_{y}, \varepsilon_{yz} = \varepsilon_{yz}^{0},$$
$$\varepsilon_{xz} = \varepsilon_{xz}^{0} + \varepsilon_{xy} = \varepsilon_{xy}^{0} + z\kappa_{xy}, \tag{4.3}$$

where plate curvatures, κ_x, κ_y, and κ_{xy} are given by

$$\kappa_{x} = -\frac{\partial \phi_{x}}{\partial x}, \quad \kappa_{y} = -\frac{\partial \phi_{y}}{\partial y}, \quad \kappa_{xy} = -\left(\frac{\partial \phi_{x}}{\partial y} + \frac{\partial \phi_{y}}{\partial x}\right). \tag{4.4}$$

For 2D modeling, the SWASS cross-section can be divided into three parts: top and bottom face sheets sandwiching a web core in the middle. The top and bottom skins are assumed to support axial and bending/shear loads, easily modeled as thin plates. The web core improves structural response by increasing bending stiffness. For simplicity, suppose that the web core mainly supports transverse shear and x-directional normal stresses. Hence, stress components distributed over the cross-sectional area can be expressed in terms of face sheets and equivalent web core. The stress–strain constitutive relation of laminate and web core for kth layer is summarized as

$$\begin{Bmatrix} \sigma_{xx} \\ \sigma_{yy} \\ \sigma_{yz} \\ \sigma_{xz} \\ \sigma_{xy} \end{Bmatrix}_{(k)} = \left[\bar{Q}_{ij}\right]_{(k)}^{\text{laminate+webequiv}} \begin{Bmatrix} \varepsilon_{xx} \\ \varepsilon_{yy} \\ \varepsilon_{yz} \\ \varepsilon_{xz} \\ \varepsilon_{xy} \end{Bmatrix}_{(k)}, \tag{4.5}$$

where $\bar{Q}_{ij}$ is an element of the transformed stiffness matrix aligned with the coordinate x and y axes. The derivation of the constitutive relationship for stress and strain may be found in [12]. Each laminate skin is assumed to be in a state of plane stress, i.e., the normal stresses along z-direction are zero. The layers are assumed to be perfectly bonded together and linearly deform with respect to rotation of the mid-plane normal. By substituting Eqs. (4.3) and (4.5) into (4.2) in conjunction with the in-plane force and moment resultants, the in-plane stiffness matrix [A], the in-plane and out-of plane coupling stiffness [B], and the bending stiffness matrix [D] are expressed as

$$A_{ij} = \sum_{k=1}^{N_k}\left(\bar{Q}_{ij}\right)_k\left(z_k - z_{k-1}\right) + A_{ij,\text{web equiv}},$$
$$B_{ij} = \frac{1}{2}\sum_{k=1}^{N_k}\left(\bar{Q}_{ij}\right)_k\left(z_k^2 - z_{k-1}^2\right) + B_{ij,\text{web equiv}}, \tag{4.6}$$
$$D_{ij} = \frac{1}{3}\sum_{k=1}^{N_k}\left(\bar{Q}_{ij}\right)_k\left(z_k^3 - z_{k-1}^3\right) + D_{ij,\text{webquiv}}, \quad i, j = 1,2,6,$$

where N_k is total number of laminate layers, z_k, z_{k-1} are the distance from the reference plane to the two surfaces of kth layer. A set of equivalent web core stiffness matrices, $A_{ij,\text{ web equiv}}$, $B_{ij,\text{ web equiv}}$, and $D_{ij,\text{ web equiv}}$, are calculated for the two configurations. The details are shown in the following sections. For transverse shear stiffness matrix,

$$K_{ij} = \sum_{k=1}^{N} \left(\bar{Q}_{ij}\right)_k \left(z_k - z_{k-1}\right), \quad i,j = 4,5. \tag{4.7}$$

The explicit expression for the transverse shear stiffness can be computed as strain energy balance under the assumption of the first-order deformation.

4.3.2.1 Equivalent Core: Volume Fraction Approach

Load is assumed evenly distributed in the web core, so that stiffness can be smeared across an equivalent single plate layer. The equivalent material properties of Young's modulus (E^*) and shear modulus (G^*) can be expressed in terms of volume fraction factor (α).

$$\begin{aligned} E^* &= \alpha E, \\ G^* &= \alpha G, \\ \alpha &= \frac{V_c - V_w}{V_c}, \end{aligned} \tag{4.8}$$

where E is the Young's modulus, G is the shear modulus, V_c is the total volume of the internal core structure space, and V_w is the total volume of web columns. Therefore, the three equivalent stiffness matrices, $A_{ij,\text{web equiv}}$, $B_{ij,\text{ web equiv}}$, and $D_{ij,\text{web equiv}}$, are computed for a single equivalent plate layer.

4.3.2.2 Equivalent Beam

In the second model, the shear webs were modeled as beams to carry concentrated in-plane and shear loads. From Timoshenko-beam theory, the coupled governing equation is given by

$$\begin{aligned} &\frac{\partial}{\partial x}\left[\kappa^2 AG\left(\frac{\partial w}{\partial x} - \psi\right)\right] + p = 0, \\ &D\frac{\partial^2 \psi}{\partial x^2} + \kappa^2 AG\left(\frac{\partial}{\partial x} - \psi\right) = 0, \end{aligned} \tag{4.9}$$

where $w = w(x)$ is the transverse deflection, $\psi = \psi(x)$ is the bending slope, κ^2 is the shape factor, G is the shear modulus, $p = p(x)$ is externally applied force along the length of the beam, and D is the bending stiffness. In this approach, the coupling matrix, $B_{ij,\text{ web equiv}}$, is disregarded, since the neutral axis of beam stiffeners is set to the location of the neutral plane of the plates. Therefore, $A_{ij,\text{ web equiv}}$ and $D_{ij,\text{ web equiv}}$ are decoupled and modeled as beams.

4.3.2.3 Smeared Slot Model

A smeared slot equivalent plate model employed volume fraction for the plate layer with slots in a similar fashion to the equivalent core model. Equivalent smeared stiffness for the slotted plate layer was expressed in terms of volume fraction factor (α) as shown in Eq. (4.8). The difference is that V_f is the total volume of the whole top face plate, and V_s the total volume of slots.

$$\alpha = \frac{V_f - V_s}{V_f} \tag{4.10}$$

4.3.3 Model Verification

In order to verify the accuracy of the equivalent smeared stiffness approach, finite element method (FEM) modeling was performed in order to see the correlation with analytical approximation for Concept 1. Length, a, and width, b, are ten times greater than height, h for this thin plate model. The NASTRAN-based FEMAP software [13] was used to model the composite layers with the equivalent core structures. QUAD4 shell elements were used for the laminate skin plates [14]. The detailed material properties are given in Table 4.1.

Figure 4.5 demonstrates the non-dimensional buckling load of the FEM equivalent model along with theoretical results as a function of aspect ratio (a/b = length/width) under compressive uniaxial load with four simply supported edges. The two equivalent configurations, which are equivalent core and equivalent beam stiffeners, are well matched. The results, comparing analytical model and equivalent models, are seen to be within 8% error at the half sinusoidal mode ($m = 1, n = 1$). The differences increase at the higher (non-critical) modes. At all three aspect ratios ($a/b = 1$, 2, and 3), the FEM critical buckling load values are closer to the analytical critical value than are the FEM higher buckling load values compared to their respective analytical higher buckling load values.

Figure 4.6 depicts the trend of buckling load as slot size varies. The figure of merit of normalized buckling load has been plotted as a function of volume fraction factor (α). It is easily observed from both equivalent models that, as the volume of slots increase, because of the reduction of stiffness, the buckling load decreases.

Table 4.1 Material properties [1]

	CFRP	Glass fiber	Copper
$E1$ ($\times 10^6$ psi) (GPa)	20.1 (138.6)	7.8 (54.8)	18.0 (124.1)
$E2$ ($\times 10^6$ psi) (GPa)	1.2 (8.3)	2.6 (18.0)	18.0 (124.1)
$G12$ ($\times 10^6$ psi) (GPa)	0.6 (4.1)	1.3 (9.0)	6.72 (46.3)
$G23$ ($\times 10^6$ psi) (GPa)	0.72 (5.0)	0.5 (3.4)	6.72 (46.3)
$G13$ ($\times 10^6$ psi) GPa	0.72 (5.0)	1.3 (9.0)	6.72 (46.3)
υ_{12}	0.26	0.25	0.34
ρ (lb/in^3) (g/mm^3)	0.0659 (0.0018)	0.0686 (0.0019)	0.322 (0.0089)

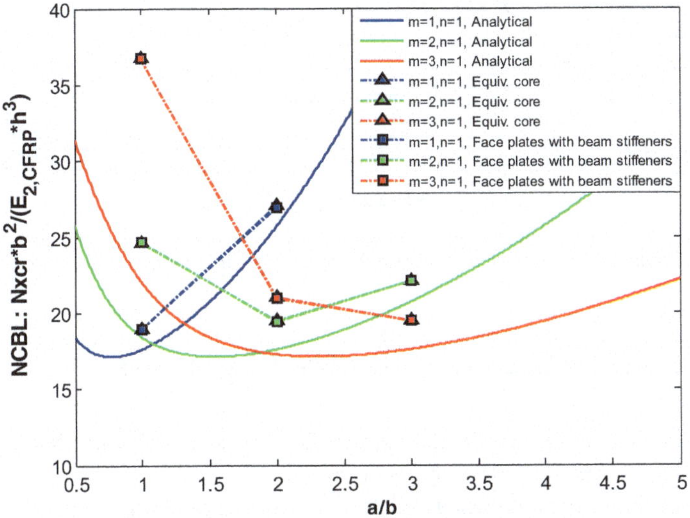

Fig. 4.5 Comparison of volume fraction method and theoretical approach without slots

However, there are significant differences between the equivalent models compared to slots modeled in the mesh (FEM) at $\alpha = 0.9$. Clearly, there is considerable approximation involved in this approach to smearing the slots, compared with a three-dimensional FEM model.

4.4 3D FEM Plate Model

The buckling response of a 3D plate model was evaluated through comparison with a previous solid FEM model studied by Sabat and Palazotto [3] and extended to the application of the four SWASS design concepts.

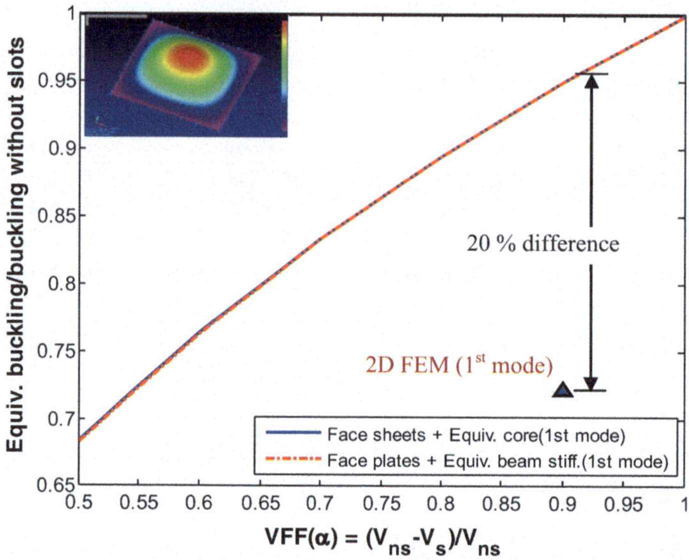

Fig. 4.6 Slot volume fraction effect: four simply supported edges under uniaxial compressive load

4.4.1 Model Verification of 3D Plate FEM Model

Our proposed 3D plate model was compared with previous work that simulates a single rectangular tube CFRP composite structure [3].

Table 4.2 shows the simulation result of the critical buckling load for non-slotted and slotted single waveguide tubes. As can be seen from the table, the difference of buckling load between Sabat's and 3D plate models is less than 3%. The 3D plate FEM model matched Sabat's 3D solid FEM model well, whereas the 3D plate model has significantly fewer number of elements than Sabat's solid model. The important conclusion to draw from the table is that the 3D plate model is more efficient than Sabat's model since it saves computation cost and is relatively accurate. However, neither simulation results match experimental results well. Further investigation is required to explain the discrepancy.

4.4.2 Structural Instability of Four Design Concepts of SWASS

AFRL recently provided four novel design concepts of SWASS for parameterization with varying panel dimensions and boundary conditions for prediction of buckling and failure. The coordinate and loading conditions for three-tube waveguide are illustrated in Fig. 4.7.

Table 4.2 Comparison of buckling load (lb_f) between Sabat's and 3D plate models

	Exp. [3]	Sabat [3]	3D plate
Non-slotted (first mode)	730.0	938.7 (28.6%)	928.9 (27.4%)
Slotted (first mode)	640.0	863.3 (35%)	892.2 (39.4%)
# of elements	N/A	40,000 (approx.)	1000 (approx.)

Difference in the parentheses () is between FEM and experiment: |FEM – Exp.|/Exp × 100 (%)

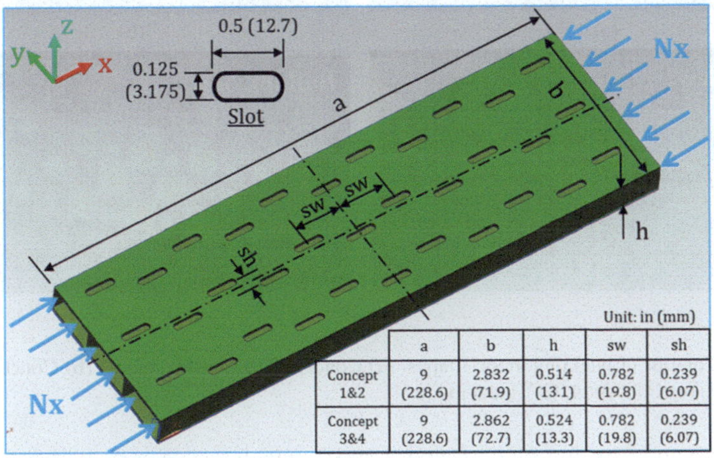

Fig. 4.7 Configuration of three waveguide tubes in SWASS sandwich construction panel under uniaxial loading

Contour plots of buckling mode shape deformation are shown in Fig. 4.8. The transverse deflection pattern of the simply supported (SS) case looks like Euler beam-column behavior: a half sinusoid wave as shown Fig. 4.8a. However, another pattern of the deflections for the clamped-clamped (CC) case are illustrated in Fig. 4.8b. Many sinusoidal waves are seen in top face sheet layers (we call it a local mode) that are structurally weakened by slots. One interesting result is shown in Fig. 4.8c where local buckling was found at the bottom panel in Concept 4, rather than on top, even though top face sheet layers are considered to be softer stiffness. This indicates that bottom face sheet layers carry more load, when the uniaxial compressive load is applied, and resulted in their buckling first.

Figure 4.9 shows the buckling comparison of a 3D plate model for the waveguide structures. These are three-tube slotted waveguides with simply supported (SS) and clamped-clamped (CC) boundary conditions on both ends, under distributed uniaxial compressive load on the cross-section of the waveguide tube. Critical buckling is normalized by the first global buckling of Concept 1 with the SS boundary condition. The SS boundary condition on both ends may not be realistic, but is useful as a lower bound for a more conservative buckling load than the CC boundary condition on both ends. The local buckling loads are approximately 1.5 times higher when slotted waveguides are designed with thin metallic foil tube with fiber skins

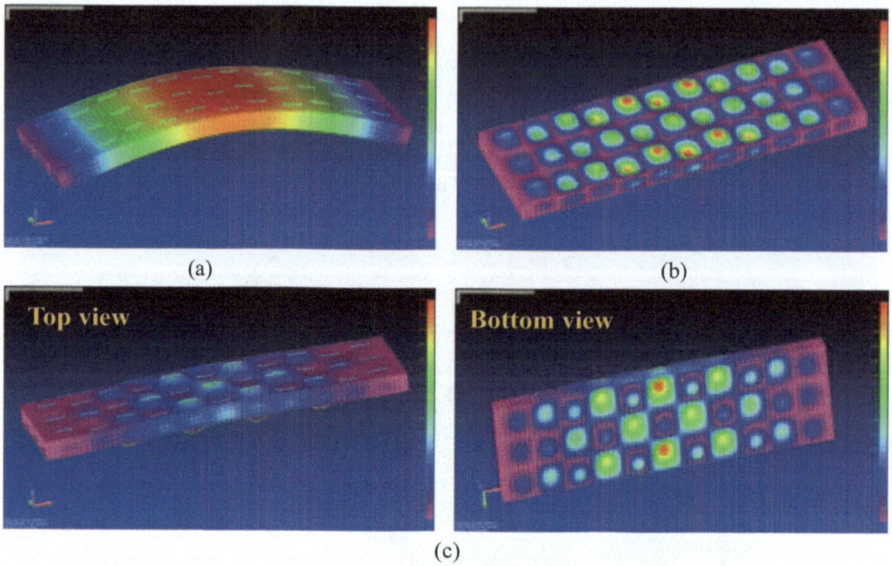

Fig. 4.8 Contour of buckling mode shapes. (**a**) Concept 2, SS, global mode. (**b**) Concept 3, CC, local mode. (**c**) Concept 4, CC, local mode

(Concept 3 and Concept 4). Concepts 1 and 2, which have a carbon fiber tube with composite laminate skins, in contrast, have a low critical buckling load.

Another plot of the non-dimensional, uniaxial buckling load normalized by weight for the four waveguide concepts is shown in Fig. 4.10. On the vertical axis, the weight-normalized buckling load is plotted where normalized buckling factor is divided by the weight of Concept 1. Three curves are shown, corresponding to the normalized buckling factors in Fig. 4.9. Most importantly, this figure provides insight into buckling variation accounting for mass ratio and provides a lightweight design criterion. In local buckling, the thin metallic foil composite waveguides have high stiffness, as well as weight efficiency, even though they are composed of metal foil having high mass density. For global buckling, Concept 1 has good mass efficiency because it is a composite structure and the lightest of all four concepts.

4.5 Structural Analysis and Evaluation

4.5.1 Nonlinear Analysis

The three-tube waveguides in a SWASS sandwich construction panel with composite face sheets were modeled for FEM nonlinear analysis [13, 14]. A variety of boundary conditions were considered. Figure 4.11a shows a fully clamped boundary condition (fully CC) on both ends where all six degrees of freedom are fixed. As

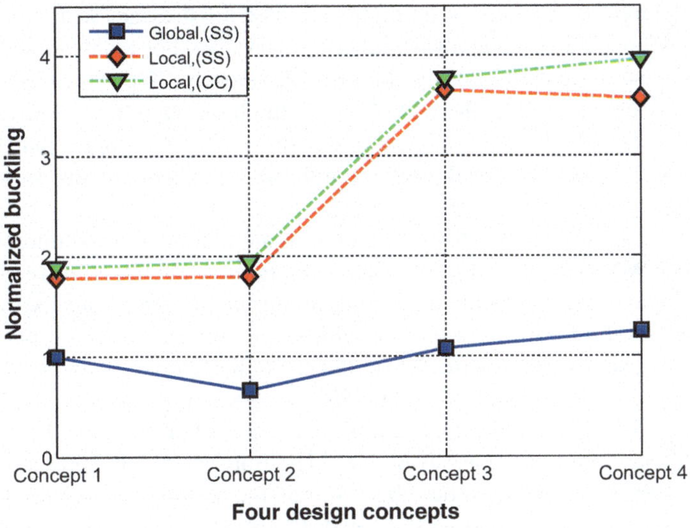

Fig. 4.9 Buckling response of three-tube slotted waveguides, where normalized buckling factor is buckling load/(first global buckling load (SS) at Concept 1)

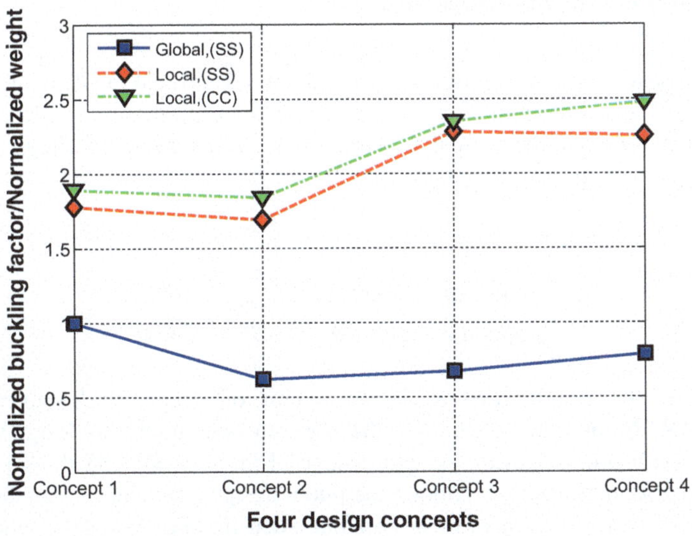

Fig. 4.10 Weight-normalized buckling of three-tube slotted waveguides, where normalized buckling factor = buckling load/(first global buckling load (SS) at Concept 1), and normalized weight = weight at each concept/(weight at Concept 1)

shown in the figure, end corners are highly strained (or stressed) with this boundary condition, which makes the analysis rather conservative. For practical design, these zones should be mitigated. Loosely clamped BC (loosely CC), shown in Fig. 4.11b, is less strictly clamped BC—the wide (y-axis) and transverse (z-axis) directions are mostly free—thereby reducing concentrated stress or strain zones near the edges and corners. Loosely clamped ends eliminated local end effects from driving the design.

Figure 4.12 shows strain in fiber directions as a function of the compressive loads for the four SWASS design concepts. Strains from the FEM model were obtained at the outer layer surface where maximum strains in the fibers are expected. The dashed lines in the figure represent the ultimate strain limits. AFRL provided the allowable fiber strain criterion of 4000 μ-strain. For the applied design load, Concept 4 is considered in more detail, since buckling analysis in previous work [4] showed that Concept 4 had the highest structural resistance when the metallic waveguide supported loads (load-bearing waveguide). The transverse deflection curve (circle marker) in Fig. 4.12d indicates the fiber strain in the region of maximum transverse deflection, which is associated with the path for buckling. Through nonlinear load-deflection analysis, the fiber strain reached the allowable strain prior to (local) buckling. That is, the allowable strain constraint prevents buckling from becoming critical. The strain limit corresponds to 12,000 lb_f (53.4 kN) compressive load—the maximum design load used for optimization.

4.5.2 Structural Optimization

In SWASS structural optimization weight reduction is critical, while achieving sufficient stiffness and strength to bear loads. The failure criteria considered are the design strain limit and the critical buckling mode, as determined through nonlinear analysis as shown in the previous section.

$$\begin{aligned} &\min_{d_S} W = f\left(d_S\right) \\ &\text{s.t.}\quad \text{design strain limit: } \varepsilon_f \le \varepsilon_{f,\text{allow}} = 4000\,\mu\varepsilon, \\ &\qquad\ \text{applied design load: } P_{cr} \le P_x = 12{,}000\,\text{lb}_f \end{aligned} \tag{4.11}$$

where W is the weight, and d_S is the vector of structural design variables. The NASTRAN optimizer was used with composite layer thickness design variables.

Figure 4.13 summarizes the weight change of the four SWASS design concepts before and after optimization. Among the initial designs, Concept 1 was the lightest composite structure because of the lowest mass density of carbon fiber, whereas due to the copper metallic foil, Concepts 3 and 4 were the heaviest. The diamond and circle markers represent post-optimization curves with different boundary conditions. The diamond marker corresponds to the fully CC on both ends and the circle marker, loosely CC. They affect post-optimization weight because the highly concentrated

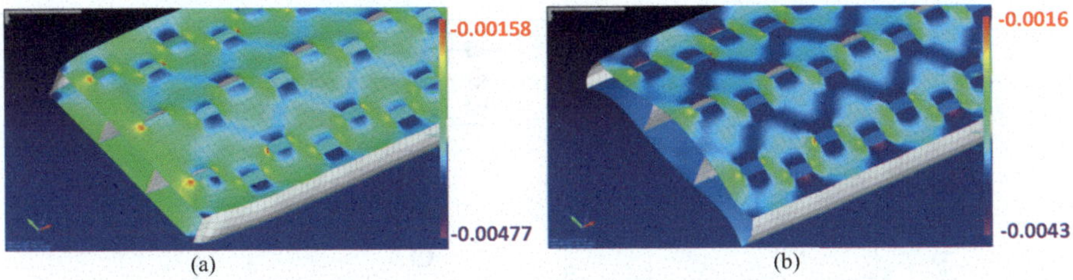

Fig. 4.11 Two different boundary conditions: contours on top and bottom face sheet strains in fiber direction (Concept 1). (**a**) Fully clamped. (**b**) Loosely clamped. Unit: in./in.

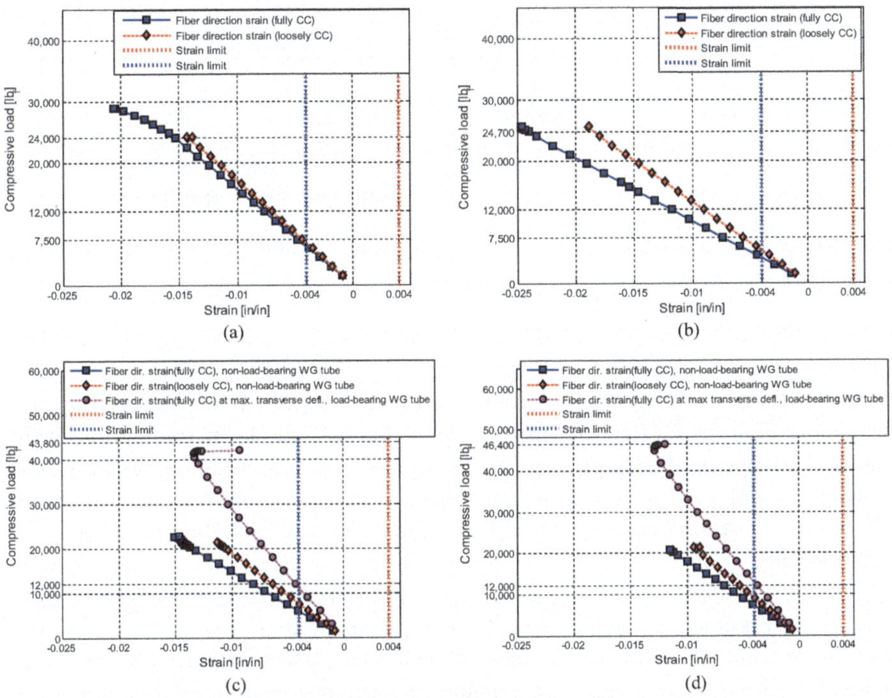

Fig. 4.12 Fiber direction strain under compressive loading in the initial design. The applied design load (12,000 lbf (53.4 kN)) is determined by the circular marker line of Concept 4. 1000 lb_f = 4.45 kN. (**a**) Concept 1. (**b**) Concept 2. (**c**) Concept 3. (**d**) Concept 4

load at the corners of the fully CC case makes layer thicknesses increase more than the loosely CC case after optimization.

Concepts 1 and 2 increased the overall composite layer thicknesses in order to satisfy the ultimate strain design criterion after optimization, resulting in a weight increase. Due to the large structural stiffening of CFRP, Concept 1 had small weight variation, compared to Concept 2. There are two reasons for the highest weight increase rate (or laminate thickness increase rate) in Concept 2 among the four

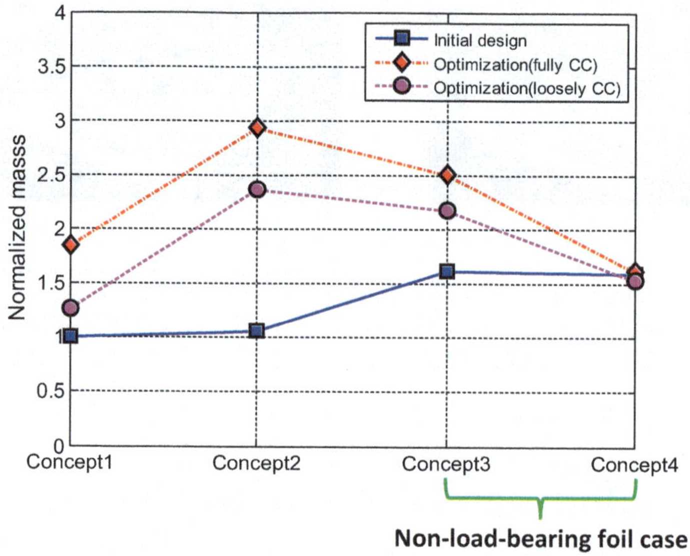

Fig. 4.13 Weight efficiency for four different SWASS design concepts before and after optimization

concepts. First, the glass fiber layers in the top and bottom skins weaken the stiffness more than the CFRP does in Concept 1. Second, the layer orientation of the waveguide tube in Concept 2 is ±45°, versus ±30° (waveguide wrapping) in Concepts 3 and 4. Load was distributed more over the top and bottom face sheets. Therefore, the strain becomes larger there, leading to increased thickness to prevent strain failure.

For Concepts 3 and 4, it was assumed in the initial structural analysis that the metallic waveguide foil carried load. However, for an optimal design, it was assumed that the metallic foil will not be adhered to the skin and therefore does not carry loads, in order to avoid degrading antenna performance from foil deflection. Thus, Concept 3 increased the weight, since the composite material carries the entire load, even though the strain limit at the applied design load is almost the same as the ultimate strain design limit at the initial design. Concept 4 has little weight variation relative to Concept 3 because the bottom skin structure (carbon fiber) has higher stiffness than the top skin (fiber glass). Therefore, it carries load more efficiently, becoming thicker, compensating for the relatively thinner top skin.

Figure 4.14 illustrates the nonlinear load-strain responses after optimization, which meets the strain and load design criteria. Concepts 1 and 4 not only have consistent resistance to load and strain, but also little variation of mass. This implies that glass fiber-based structures may be screened and CFRP-based or mixed structures with CFRP can be used.

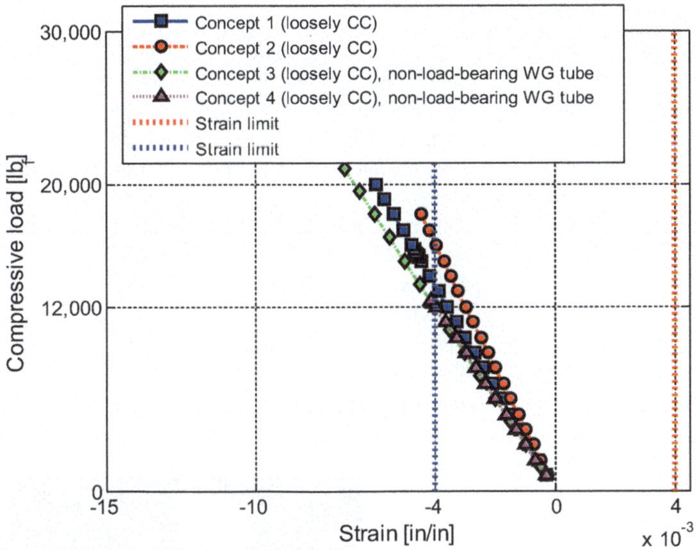

Fig. 4.14 Post-optimization curves

4.5.3 *Experimental Results*

The specimens corresponding to Concepts 3 and 4 were chosen for comparison with tests. Concept 4 specimens use Fibreglast style 7718 glass fiber weave [15] for top skin, Fibreglast style 660 carbon weave for bottom skin. The 0.005-in. (0.127 mm)-thick waveguide tube is wrapped by Fibreglast 2610 glass biaxial glass fiber. Bottom skin of Concept 3 specimens use Fibreglast style 7718 glass fiber weave. The overall sizes and lay-up orientations are shown in Table 4.3. For simulations, we assumed that composites are perfectly bonded, and woven fabrics are stacked layer by layer. The general material properties used are shown in Table 4.1 since the manufacturer does not provide material properties.

The compressive load test was done using an MTS machine as shown in Fig. 4.15. Test device and fixtures followed ASTM C364 to measure load-bearing capacity. Strain gages were placed at the center point of the outer surface on both sides for data collection in the longitudinal direction, that is, compressive loading direction. Results for two sets of boundary conditions are shown: clamped and simply supported conditions on both ends. The silver cylinder clamped the test panels, so the panels cannot experience rotations and deflections at both ends. The silver cylinder and black jig line contacted. Therefore, it allowed the panels to rotate about the cylinder center axis, although there is friction. For simulation, we considered two extreme cases, simply supported and clamped, because these two simple conditions are expected to bound the results for the actual boundary conditions.

Table 4.3 SWASS compression test panel

ID	Bottom skin (SG1)	Top skin (SG2)	Tube wrap	Length (in mm)	Width (in mm)	Total height (in mm)
6	5 ply[a] 660 CFRP	5 ply E-glass	2 ply biaxial fiberglass	2.12 (53.8)	3.129 (79.5)	0.435 (11.0)
7	5 ply 660 CFRP	5 ply E-glass	2 ply biaxial fiberglass	2.15 (54.6)	3.134 (79.6)	0.435 (11.0)
9	5 ply 660 CFRP	5 ply E-glass	2 ply biaxial fiberglass	2.13 (54.1)	3.088 (78.4)	0.435 (11.0)
8	5 ply E-glass	5 ply E-glass	2 ply biaxial fiberglass	2.088 (53.0)	3.160 (80.2)	0.428 (10.9)
10	5 ply E-glass	5 ply E-glass	2 ply biaxial fiberglass	2.1 (53.3)	3.137 (79.7)	0.424 (10.8)

[a]All 5 ply laminates are 0/90, ±45, 90/0, ±45, 0/90

Fig. 4.15 MTS compressive load test (ASTM C364)

Figure 4.16 shows load-strain curves corresponding to Concept 4. Figure 4.16a, which is the glass fiber top skin, shows that experiment results differ little from the simply supported FEM simulation. This confirms that the experiment was predominantly simply supported, as intended. Figure 4.16b for CFRP bottom skin is not conclusive, because the experimental results were nonlinear. Fig. 4.17 corresponds to Concept 3. The simulation curves between fully CC and SS-SS BCs are almost the same because the structures are symmetric about top and bottom skins, but failure occurs first for the SS-SS BC, because it allows rotations. The simulation results overlay the experiment results except one, demonstrating the accuracy. Boundary conditions of the experiments was suspected as the cause of the nonlinearity.

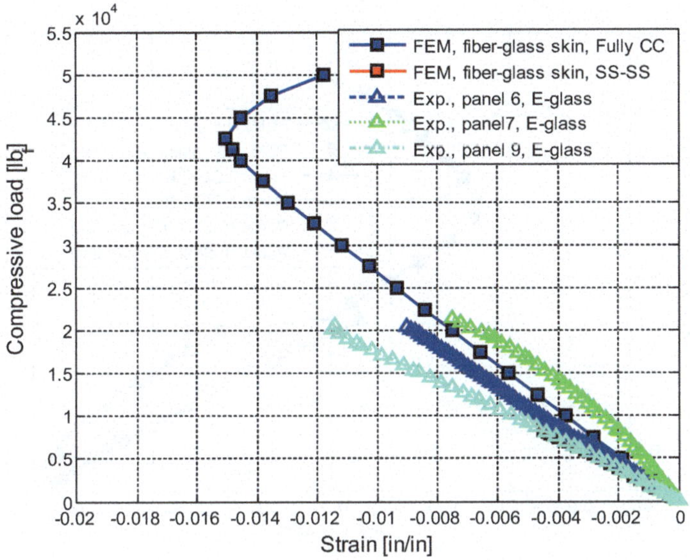

(a)

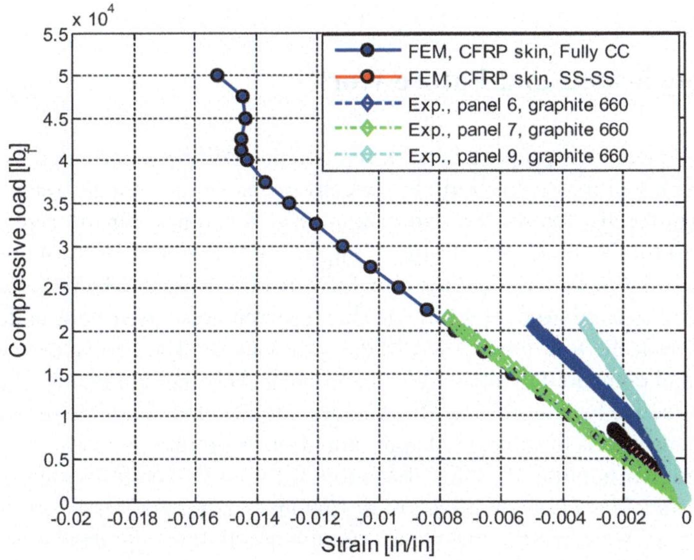

(b)

Fig. 4.16 Comparison of fiber direction strains: 5ply E-glass on both sides (1000 lb_f = 4.45 kN). (**a**) E-glass top skin. (**b**) CFRP bottom skin

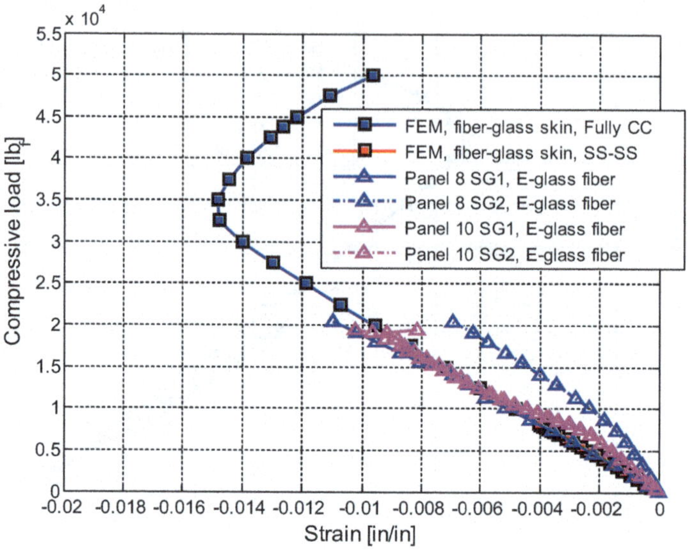

Fig. 4.17 Comparison of fiber direction strains: 5ply E-glass on both sides (1000 lb_f = 4.45 kN)

4.6 Conclusions and Future Work

Our research team has proposed and developed the SWASS, configured as composite waveguide tubes, covered by electromagnetically transparent or non-electromagnetically transparent laminate skins. The tubes can be replaced with metallic foil for the alternative design to improve EM radiation performance. These composite structures are critical not only for structural weight to be reduced but also for structural strength to be reinforced. This research presents structural design and optimization of the four novel SWASS design concepts. The equivalent model was proposed and compared to an analytical approach. This method is important in that it is cost-effective by simplifying 3D structure into 2D plate model to indicate rough trends, but also leads to substantial approximation error when smearing the effect of slots. For higher fidelity, 3D plate FEM models for these structures are proposed to achieve sufficient accuracy with the most savings in computational cost. Nonlinear strain analysis evaluated the mechanical failure and lightweight design criteria and provided optimal design solution. It was shown that Concept 1 (CFRP WG) and Concept 4 (CFRP + Copper WG) are good. Concept 4 may be best in both structural and EM efficiencies (subject to further EM performance analysis). The 3D plate modeling was validated against experiments. They were in good agreement in some cases, but there are still a few uncertainties such as boundary conditions and bonding strength of layers. In summary, the three main contributions are (1) verification that at least a slotted plate model is required for sufficient accuracy compared to a smeared slot stiffness model; (2) design performance of four novel SWASS concepts was quantified; and (3) slotted plate FEM models of the novel SWASS concepts were validated against experimental results. For future work, we propose research that will continue to optimize simultaneously the structural and electromagnetic performance.

Acknowledgements This material is based on research sponsored by the Air Force Research Laboratory under the agreement number FA8650-09-2-3938. The US Government is authorized to reproduce and distribute reprints for governmental purposes notwithstanding any copyright notation thereon.

References

1. Callus PJ (2007) Conformal load-bearing antenna structure for Australian Defence Force Aircraft, Australia, 50p
2. Callus PJ (2008) Novel concepts for conformal load-bearing antenna structure. Defense Science and Technology Organization, Aeronautical and Maritime Research Laboratory, Australia, p 94
3. Sabat JW, Palazotto AN (2013) Structural performance of composite material for a slotted waveguide antenna stiffened structure under compression. Compos Struct 97:202–210
4. Kim W, Canfield RA, Baron W, Tuss J, Miller J (2013) Modeling and simulation of slotted waveguide antenna stiffened structures. In: 19th international conference on composite materials, Montreal, Canada
5. Ha T, Canfield RA (2011) Design optimization of a WR-90 slotted waveguide antenna stiffened structures. In: 52nd AIAA/ASME/ASCE/AHS/ASC structures, structural dynamics and materials conference, April 4–7. American Institute of Aeronautics and Astronautics Inc., Denver
6. Kim W, Ha T, Canfield RA, Baron W, Tuss J (2014) Radio frequency optimization of slotted waveguide antenna stiffened structure. Submitted to Aerospace Science and Technology
7. Smallwood BP, Canfield RA, Terzuoli AJ Jr (2003) Structurally integrated antennas on a joined-wing aircraft. In: 44th structural dynamics, and materials conference. American Inst. Aeronautics and Astronautics Inc., Norfolk
8. Knutsson L, Brunzell S, Magnusson H (1985) Mechanical design and evaluation of a slotted CFRP waveguide antenna. In: Fifth international conference on composite materials, ICCM-V. Metallurgical Soc Inc., San Diego
9. Timoshenko S, Woinowsky-Krieger S (1959) Theory of plates and shells. McGraw-Hill, New York
10. Gürdal Z, Haftka RT, Hajela P (1999) Design and optimization of laminated composite materials. Wiley, New York
11. Palazotto ANDST (1992) Nonlinear analysis of shell structures. American Institute of Aeronautics and Astronautics, Washington, DC
12. Reddy JN (2004) Mechanics of laminated composite plates and shells: theory and analysis, 2nd edn. CRC Press, Boca Raton
13. FEMAP, FEMAP 10.3. 2011, Siemens Inc.
14. MSC.NASTRAN, MSC Nastran Quick Reference Guide 2012, in 2012, MSC Software
15. Fibre Glast Development Corporation. http://www.fibreglast.com/

Chapter 5
The Influence of Azobenzene Content on Azopolyimides Capacity to Form Laser-Induced Surface Relief Gratings

Ion Sava and Iuliana Stoica

5.1 Introduction

Azobenzene is the simplest aromatic azo compound and a primary structure of many derivatives. Azo group (–N=N–) is very important due to its ability to absorb light in specific wavelength ranges and are attached to as azo chromophores. A great interest has been made in the development of materials that are optically responsive and have the potential to be useful in many photonic technologies. Polymers which contain azobenzene (azopolymers) and many of its derivatives are promising materials for such applications. Owing to the existence of the aromatic azo cores, azopolymers usually show strong light absorption in the ultraviolet (UV) and/or visible spectral range. The energy acquired from the absorbed light will initiate various photoresponsive variations in properties. By introduction of different numbers of azo linkages and the types of rings and substituents, it is possible to synthesize a lot of aromatic azopolymers possessing different properties useful in various applications [1].

The properties of azopolymers are dependent on the azo linkage itself but also are controlled by the groups on both sides of the bridge. The azo groups can be covalently bonded to benzene rings, naphthalene rings, aromatic heterocycles, and other aromatic rings. For azopolymers, the strong absorptions appearing in above wavelength range are mainly related to the electronic transitions of the aromatic azo core structures. Due to the light absorption, the electronic excitation can cause energetic, structural, and dynamic variations. One of the most important variations at the molecular level is the *trans–cis* photoisomerization of the azo cores. Reversible *trans–cis–trans* isomerization of the molecules causes structural changes resulting also in spectral changes. During *trans* to *cis* conversion, the intensity of the absorption band corresponding to the n–π* transition increases, while the π–π* band

I. Sava (✉) · I. Stoica
"Petru Poni" Institute of Macromolecular Chemistry, Iasi, Romania
e-mail: isava@icmpp.ro

C. Miron et al. (eds.), *Carbon-Related Materials*,
https://doi.org/10.1007/978-3-030-44230-9_5

decreases [2]. Structural changes of the azobenzene are reflected in obtaining a stable dipole moment by the *cis*-isomer (3.0 D) and in the change in geometry caused by a reduction in the distance between carbon atoms in positions 4 and 4′ from 9.0 Å for the *trans* isomer to 5.5 Å for the *cis*-isomer [2, 3].

The *trans–cis* photoisomerization is one of the most important properties of aromatic azo compounds, which has been widely explored to develop azopolymers with various photoresponsive functions. Owing to the simplicity, azobenzene and its derivatives have been intensively studied as extremely systems to understand the mechanism of the photoisomerization. Unfortunately, the photoisomerization mechanism of aromatic azo compounds have not been fully clarified because most of the experimental results are not arguable and their interpretations are often subjects of heated debates [4–7]. The photoisomerization rate and degree of the azo chromophores depend on the factors such as free volume, temperature, mobility, and polarity of the polymer systems [3, 8–12].

In the past decades, azopolymers with diverse compositions and structures have been synthesized. The azo functional groups can be incorporated into the polymer main chain or, much more frequently, be linked to the backbone as side-chain pendant groups either directly or via a spacer group. There are two typical ways to synthesize azopolymers: (a) polymerization and copolymerization of monomers that contain azo functional groups and (b) chemical modification of a suitable precursor polymer to introduce the azo functional groups after polymerization [2, 13–20].

Linearly polarized light orients the azo chromophores perpendicular to the polarization direction of the incident light. The chromophore orientation is caused by *trans–cis–trans* isomerization cycles under light absorption. Due to this preferential orientation, the light irradiation will cause the macroscopic dichroism and birefringence, which are termed photoinduced dichroism and birefringence. The photoinduced dichroism and birefringence effects have been widely explored for the applications in optical gratings and holography. A simple mask exposure technique can be used for the grating fabrication. Nowadays, two-beam interference technique is more widely used to produce gratings through exposure with spatially modulated light field [1–3, 21]. Upon irradiation a sinusoidal surface patterns can be developed on the azopolymer films at temperature well below the glass transition temperature (T_g) of the polymers. The gratings can be erased by heating the samples above their T_{gs} without degradation or charring of the polymer films. The SRG formation caused by the mass transport has drawn wide attention not only because of the mazy nature of this physical effect but also owing to potential applications as a new approach to fabricate various optical elements, functional surfaces, and devices [3, 4]. The terms "surface gratings" and surface relief gratings (SRG) were first used by the discoverers of this unique photoresponsive effect [22, 23]. Besides using a continuous wave (cw) laser as the light source in most investigations, mode-locked lasers with the pulse of very short duration have also been used to inscribe SRGs [24]. In this way, a frequency-doubled yttrium–aluminum–garnet (YAG) laser operating at 532 nm is typically used as the light source, which has pulses of 5–7 ns duration at a 20 Hz rate [25–28]. The used methods to characterize the topographic morphology and the

SRG formation process are atomic force microscopy (AFM) and diffraction efficiency measurement. Atomic force microscopy (AFM) is most widely used to investigate the SRGs, which can give the information about the grating parameters and surface profile [1]. The SRG formation behavior is significantly affected by the polymer structures and optical parameters. Optical parameters of the writing laser beam like wavelength, intensity, polarization, spot size, and intensity distribution have important effects on the SRG formation. As SRGs are inscribed by two-beam interference, the intersection angle is an important parameter to determine the grating periods. The first process is the light-induced orientation of azobenzene moieties perpendicular to the polarization plane of the incident light. The second process is the mass transport along the grating vector to form SRG. It is well known that the segments of a polymer chain can undergo a migration in considerable scale only when temperature is above T_g of the polymer. The SRG formation, as an uncommon phenomenon, seems to be against both instinct and conventional belief. The first explanation at the early stage investigation was that the SRG formation is a result of mass diffusion of the polymer chains caused by the temperature gradient, but after much more investigations it was established that the large amplitude SRGs can only be formed on polymer films that contain a substantial amount of photoisomerizable azo chromophores [1]. The thermal effects play a negligible role during SRG formation for films of any thickness [29–34].

In the case of photoinduced anisotropy or surface relief gratings, all the light generated effects are reversible. The isotropy is restored upon heating the sample or using a light beam of a circular or linear polarization of a certain direction [1, 7, 35]. For practical applications, the azopolymers should fulfill other requirements like short response time, long-term chemical and orientational stabilities at the operating temperatures as well as good film forming ability, high thermal stability, and good processability. The physical and photoinduced properties are strongly related to both the chromophore and polymer matrix architecture. Our approach for increasing the stability of optically induced birefringence refers to the utilization of high glass transition temperature (T_g) polymers, such as polyimides. Aromatic polyimides belong to a family of high-performance polymers and have played a key role in the materials for the microelectronics, optics, and aerospace industries due to their superior thermal stability, chemical resistance, and mechanical durability [36–41]. A lot of researches are dedicated to the synthesis and characterization of aromatic polyimides which contain azobenzene group (azopolyimides) [42–47].

This chapter tries to concentrate our previous studies dedicated to SRG formation using pulse laser irradiation technique of some azopolyimides and azocopolyimides. To obtain the surface relief gratings, it was used a radiation corresponding to the third harmonic of a pulsed Nd:YAG laser (Brilliant B from Quantel, 355 nm wavelength, 6 ns pulse duration, 6 mm in diameter) to generate an interference field in the near-field of a phase mask (1000 grooves per mm). The used energy of incident fluency was in the range of 8–45 mJ/cm^2. The number of laser pulses was between 1 and 100. The atomic force microscopy was used to characterize the topographic morphology of the SRG.

5.2 The Synthesis of Azopolyimides and Azo-copolyimides

Two types of azobenzene containing diamines (azodiamines) have been used in the synthesis of azopolyimides. The first type of aromatic diamines which contain azobenzene groups have been synthesized by diazotation of *ortho-*, *para*-toluidine, or *para*-chlor aniline followed by coupling with *meta*-phenylene diamine [15]. These azodiamines have been used for the synthesis of azopolyimides and azo-copolyimides. The synthesis of diamines containing azobenzene group is presented in Scheme 5.1.

The second type of azodiamines have been synthesized by a multistep reaction which starts with diazotation of *para*-toluidine, coupling with phenol and then the reaction of the resulted *p*-hydroxy-*p′*-methyl-azobenzene with dibromoalkane. The *p*-bromoalkyloxy-*p′*-methylazobenzenes containing different length of alkyl chains have been used for the obtaining of azodiamines by Williamson reaction with an *ortho*-hydroxydiamine such as 2,2-bis(3-amino-4-hydroxyphenyl) hexafluoropropane. The reaction takes place at reflux, in acetone in the presence of K_2CO_3 and some traces of KI, for around 60 h (Scheme 5.2) [48].

The azopolyimides were obtained starting from aromatic dianhydride (such as benzophenonetetracarboxylic dianhydride, 4,4′-(4,4′-isopropylidenediphenoxy) bis(phthalic) anhydride or hexafluoroisopropylidene dianhydride), and aromatic diamines having azobenzene groups and various substituents on a benzene ring [49, 50]. The structure of the azopolyimides is shown in Scheme 5.3.

Three azo-copolyimides have been synthesized through the polycondensation of (hexafluoroisopropylidene)-diphthalic dianhydride and a mixture of a diamine containing preformed substituted azobenzene group, namely 2,4-diamino-4′-methylazobenzene and three aromatic diamines which contain ether linkage, such as bis(p-aminophenoxy)-1,4-benzene, bis(p-aminophenoxy)-1,3-benzene, or bis(p-aminophenoxy)-4,4′-biphenyl (Scheme 5.4). The molar ratio between the two diamines is 1:3. Details concerning the polymers synthesis and characterization were previously reported [19, 50].

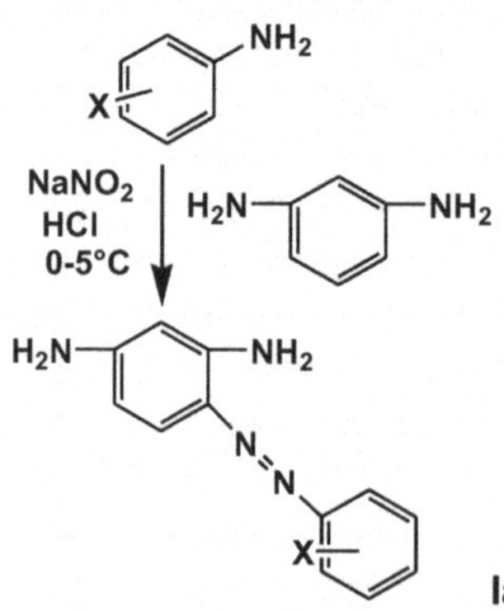

Ia: X = *o*-CH_3; Ib: X = *p*- CH_3; Ic: X = *p*-Cl

Scheme 5.1 Synthesis of aromatic diamines containing side azobenzene groups, **I**

Scheme 5.2 Synthesis of azodiamines containing two-side azobenzene groups

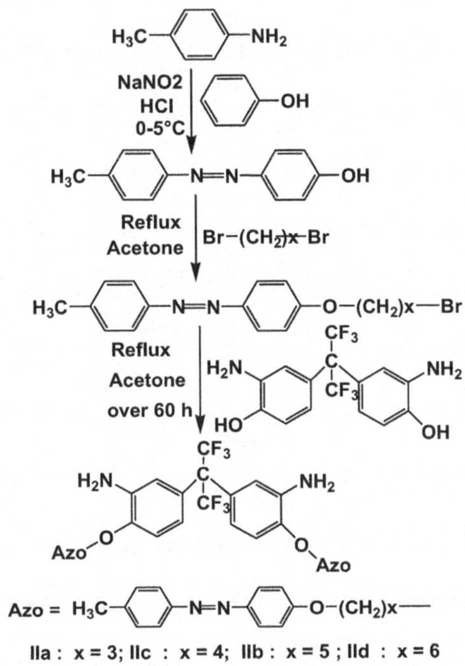

The azodiamines containing two-side azobenzene groups with different length of alkyl chains were used in the synthesis of azo-copolyimides by the polycondensation reaction of benzophenonetetracarboxylic dianhydride and a mixture of two diamines, one contains two pendent substituted azobenzene group, namely 2,2 bis-(3-amino-4-alkyloxy-phenyl-4′-methylazobenzene) hexafluoropropane and one of which contains methyl substituted phenyl groups, namely 4,4′-diamino-3,3′-dimethyldiphenylmethane (MMDA). The molar ratio between the two diamines was 0.25:0.75, respectively. The polycondensation reaction of these monomers was carried out by using a procedure reported previously [51]. The structure of the azo-copolyimides are shown in Scheme 5.5.

5.3 Photochromic Behavior

A first general observation regarding the photochromic behavior of azopolyimides is that the maximum conversion degree from *trans* to the *cis*-isomer in the solid state is lower than that of the corresponding azopolyimides in solution. Thus, a degree of isomerization within 52–69% was obtained in solution for polyimides with hexafluoroisopropylidene units, compared with 28–39.5% for polymers with isopropylidene or 49–57% for polymers with benzophenone units [15, 49, 50].

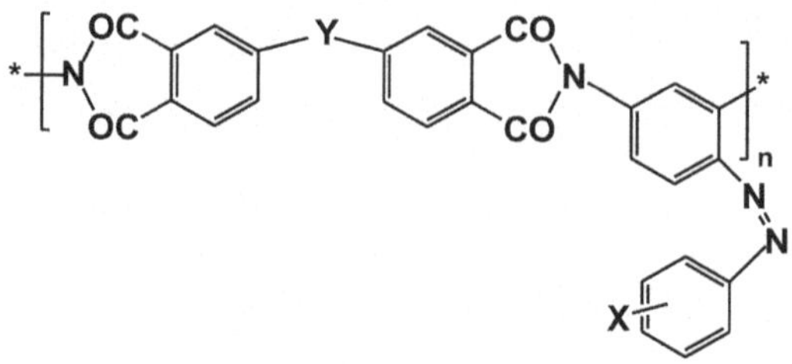

AzoPI-1: X = o-CH_3; Y = CO **AzoPI-4**:X = o-CH_3 ; Y = $C(CF_3)_2$

AzoPI-2: X = p-Cl; Y = CO **AzoPI-5**:X = p-Cl; Y = $C(CF_3)_2$

AzoPI-3:X = p-CH_3 ; Y = CO **AzoPI-6**:X = p-CH_3 ; Y = $C(CF_3)_2$

AzoPI-7:X = o-CH_3 ; Y = O – C_6H_4 - $C(CH_3)_2$– C_6H_4 -O

AzoPI-8:X = p-Cl; Y = O – C_6H_4 - $C(CH_3)_2$– C_6H_4 -O

AzoPI-9:X = p-CH_3 ; Y = O – C_6H_4 - $C(CH_3)_2$– C_6H_4 -O

Scheme 5.3 Structure of the azopolyimides **AzoPI-1–AzoPI-9**

As one can see in Fig. 5.1, the photoisomerization equilibrium is strongly influenced by the chemical structure. The higher values of conversion to *cis*-isomer, 69%, were obtained for the polymer which contains hexafluoroisopropylidene units and methyl substituent in azobenzene group in *para*-position. The methyl substituent in *ortho*-position produces some steric hindrance, and for this reason the maximum conversion degrees for these polymers were lower in comparison with azopolymers which present the similar chemical structures but with methyl substituent in *para*-position (Fig. 5.1). The isomerization process in solid state is different compared with that in solution. For all the investigated polyimides, in solid state, the maximum conversion to the *cis*-isomer is within 22.5–39%, with higher values for polymers which contain hexafluoroisopropylidene units. The reason could be the similar free volume values of azopolyimides, in spite of the differences concerning the chain rigidity and chain conformation.

The *cis–trans* relaxation in natural visible light shows that the requested time for *cis–trans* isomer conversion is 400–500 s, in the case of azopolyimides, which contain hexafluoroisopropylidene or benzophenone units, while for the other azopolyimides the conversion time is 10 times longer, being around 4000–5000 s. It can be mentioned that the higher *cis–trans* relaxation rate of the flexible polymer film is favorable to the reordering processes. This conformational instability state is a consequence of the azobenzene continuous photoisomerization which may be accompanied, in relation with the chemical structure, by strong dipole-moment fluctuations

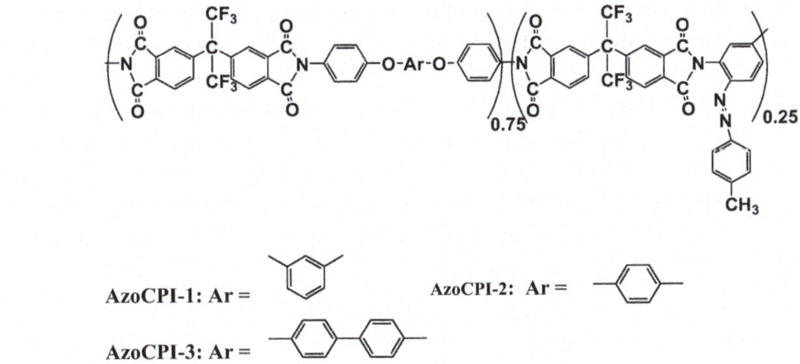

Scheme 5.4 Structure of azo-copolyimides **AzoCPI1–AzoCPI3**

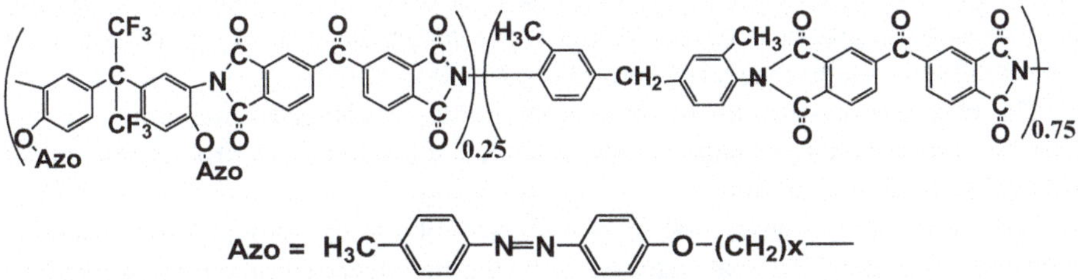

AzoCPI-4: x = 3; AzoCPI-5: x = 4; AzoCPI-6: x = 5; AzoCPI-7: x = 6

Scheme 5.5 Structure of azo-copolyimides **AzoCPI-4–AzoCPI-7**

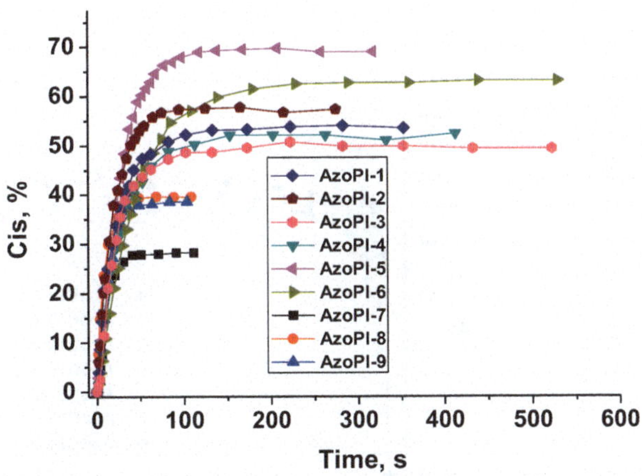

Fig. 5.1 Plot of the *cis*-form azo-group content as a function of UV irradiation time corresponding to the azopolymers **(AzoPI)** in solution ($CHCl_3$)

along the chain. This continuous motion of the azo groups induces a conformational instability in the entire polymeric chain which prevents the phase stabilization in the solid state.

The photochromic study effectuated in solid state for azo-copolyimides showed interesting results. A general observation is that the maximum conversion degree to the *cis*-isomer is situated in the range of 39–57% with slightly lower values for **AzoCPI-1**, **AzoCPI-2** and **AzoCPI-3**. The introduction of flexible alkyl units in the macromolecular chains of **AzoCPI-4–AzoCPI-7** probable facilitate the conversion to the *cis*-isomer. It is worthy to note that the isomerization degrees of the azo-copolyimides with odd number of methylene units were attain faster the isomerization degree then that the compounds with even number of methylene units. The better values concerning the maximum *cis*-isomer content were obtained for the polymer **AzoCPI-6** that can reach a 57% conversion degree (Fig. 5.2). The odd-even effect is probable due to some intermolecular steric hindrance between the azo groups and the main chains. Regarding the time necessary to attain the maximum isomerization degree was observed that is shorter in the case of compounds with odd member of methylene units being only 25–30 s [52].

The absorption spectrum of the azopolyimides evidenced a maximum within 330–346 nm wavelength range but also a tendency for higher absorption maxima at wavelengths below 250 nm.

The thermogravimetric analysis of all azopolyimides and azo-copolyimides revealed a very good thermal stability, as the thermal degradation onset temperature was above 300 °C, this step being attributed to the azo groups which are more vulnerable part of the polymers [15, 19, 20, 48, 50, 52]. All azopolyimides and azo-copolyimides showed a high glass transition temperature being in the range of 185–241 °C [15, 50–52].

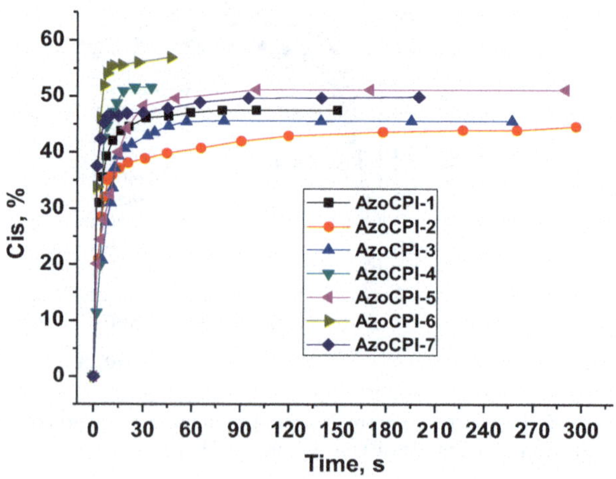

Fig. 5.2 The *cis*-isomer content vs. time for **AzoCPI-1–AzoCPI-7** in solid state

5.4 Surface Structuring Investigations

The behavior of azopolyimides and azo-copolyimides to the structuring under polarization light showed that the irradiation conditions significantly influence the surface geometry of the polymers. Surface irradiation was carried out under the action of a light field with a controlled distribution represented by an interference pattern. A great influence to the modulation depth and to the uniformity of the surface relief gratings is due to the content of the azobenzene in the macromolecular chain of the polymers. In the case of these azopolyimides with a relatively high content in azobenzene (24–32%) to obtain SRG of good quality is necessary to use a low pulses number and a high energy fluence. As a representative example in the case of azopolyimide **AzoPI-6** with a content in azo chromophore of around 28%, it was observed a modulation depth of around 350 nm by using a high energy fluence (35 mJ/cm^2) and a reduced pulses number (5 pulses). In this case, the uniformity of the surface relief was very good, the deviations being in the region of 50–60 nm (Fig. 5.3). By increasing the number of pulses, the uniformity of the surface was disturbed probably due to some supplementary modulation effect. The situation was different by using a low energy fluence and a low pulses number. The modulation depth was reduced to only 15–20 nm, no well-defined surface relief gratings was observed and the generated channels were not uniform. By increasing the number of pulses to 100, a uniform surface relief grating was obtained but the maximum modulation depth was only 100 nm.

Figure 5.4 shows the evolution of the modulation depth of the surface relief gratings of the **AzoPI-6** by increasing the pulses number and by using different energy fluence. As can be seen by using a low energy fluence (8.4 mJ/cm^2) and by increasing the pulses number until 100 the maximum modulation depth was only 90 nm. On contrary by using 35 mJ/cm^2 and only five pulses, the maximum modulation depth was around 350 nm and by increasing the number of pulses the modulation depth begins to decrease and the uniformity of the channels disappears [15, 35, 53, 54].

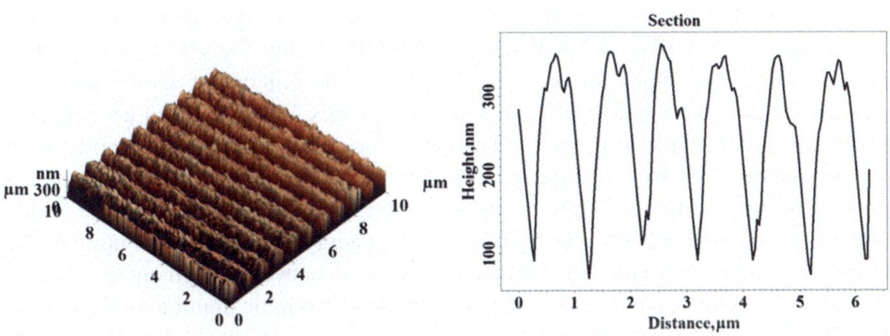

Fig. 5.3 AFM image of the structured surface and corresponding cross-section profile of azopolyimide film **AzoPI-6** after irradiation with 35 mJ/cm^2 by using five pulses

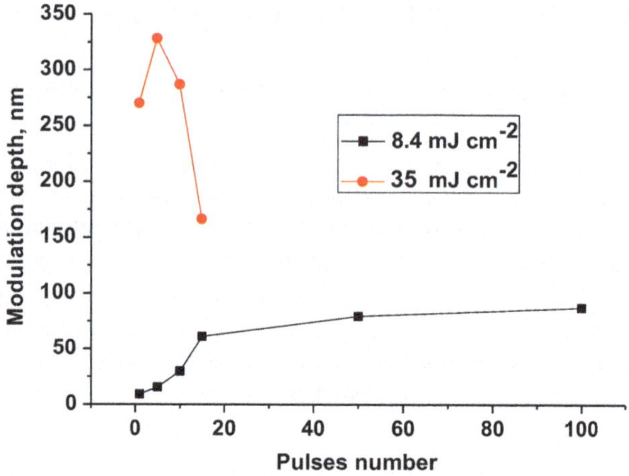

Fig. 5.4 The modulation depth vs. pulses number for azopolyimide **AzoPI-6**

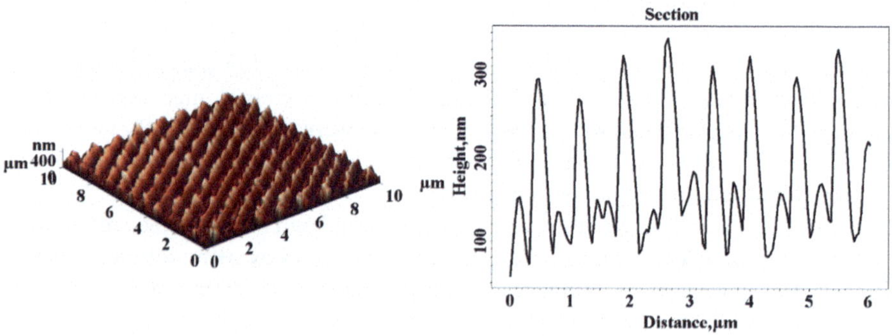

Fig. 5.5 AFM image of the structured surface and corresponding cross-section profile of azo-copolyimide film **AzoCPI-2** after irradiation with 45 mJ/cm^2 by using 100 pulses

To obtain free-standing films, we proceed to synthesized azo-copolyimides by using a mixture of aromatic diamine and an azodiamine taking in the molar ratio of 3 to 1 and 6FDA as dianhydride. In this context, we synthesized three azo-copolyimides **AzoCPI-1, AzoCPI-2,** and **AzoCPI-3**, and the free-standing films were irradiated to obtain surface relief gratings. The content of azobenzene was very low, between 1.8 and 1.99%. In spite of this very low content in azobenzene, all the azo-copolyimide films gave very nice surface relief gratings with modulation depth in the range of 30–240 nm depending of the structure of aromatic diamine and the structuring conditions. Thus, in the case of a flexible aromatic diamine which contain *meta*-catenated linkages by using a low energy and low pulses number the modulation depth of 50 nm was uniform but became uneven at high energy fluence irrespective of pulses number. By using a more rigid aromatic diamine in the case of **AzoCPI-3** irrespective of the irradiation conditions, it was obtained very well surface relief gratings with modulation depth in the domain of 80–200 nm with higher values for high energy and high pulses number [19]. A representative example for good AFM image is given in Fig. 5.5.

Table 5.1 The modulation depth and texture direction index vs. pulses number and energy fluence for azo-copolyimides **AzoCPI-1–AzoCPI-3**

Polymer	Energy (mJ/cm²)	Pulses number	Modulation depth (nm)	Stdi
AzoCPI-1	10	10	50	0.303
AzoCPI-2	10	10	30	0.529
AzoCPI-3	10	10	100	0.232
AzoCPI-1	10	100	40	0.546
AzoCPI-2	10	100	150	0.295
AzoCPI-3	10	100	100	0.252
AzoCPI-1	45	10	100	–
AzoCPI-2	45	10	100	0.197
AzoCPI-3	45	10	80	0.261
AzoCPI-1	45	100	180	0.437
AzoCPI-2	45	100	240	0.211
AzoCPI-3	45	100	200	0.195

The preferential orientation of the surface morphology was characterized by using the texture direction index (Stdi). As can be seen in the Table 5.1, surfaces with very dominant directions have Stdi values under 0.35, and those with no texture orientation have Stdi more than 0.4. By using an energy density of 10 mJ cm^{-2} and 10 pulses, the obtained surface photoinduced pattern is well defined in the case of azo-copolyimides **AzoCPI-1** and **AzoCPI-3** and uneven for polymer **AzoCPI-2**. This fact is also argued by the values of Stdi, which were 0.303 and 0.232 for **AzoCPI-1** and **AzoCPI-3,** respectively, indicating a good anisotropy of the morphology and slightly higher (0.529) for polymer **AzoCPI-2,** showing a mild texture orientation [19]. The same behavior was presented in Table 5.1 for other irradiation conditions.

Very nice free-standing films have been obtained for azo-copolyimides which have in the same structural unit two pendant azobenzene linkages, **AzoCPI-4–AzoCPI-7.** Uniform and orientated surface relief gratings were obtained for these azo-copolyimides with a maximum azobenzene content of 5%, irrespective of pulses number (10 or 100), using a laser field with an energy density of 45 mJ/cm^2. It is worth to mention that by increasing the number of methylene units the modulation depth slightly decreases being in the range of 220–140 nm by using a high energy fluence (45 mJ/cm^2) and 100 pulses. By using a low pulses number and high energy fluence, the modulation depth has the same tendency, but the observed values were slightly lower being in the domain of 134–88 nm (Fig. 5.6). The same behavior was early reported by other authors [55].

Irrespective of the irradiation condition, the texture direction index had low values, being in the range of 0.10–0.34, proving good anisotropy of the morphology of the polymer films (Table 5.2). Under the same light irradiation condition, but using low energy fluence, irrespective of used pulses number no surface relief gratings were formed. An AFM representative image of the SRG and the corresponding cross-section profile is given in Fig. 5.7.

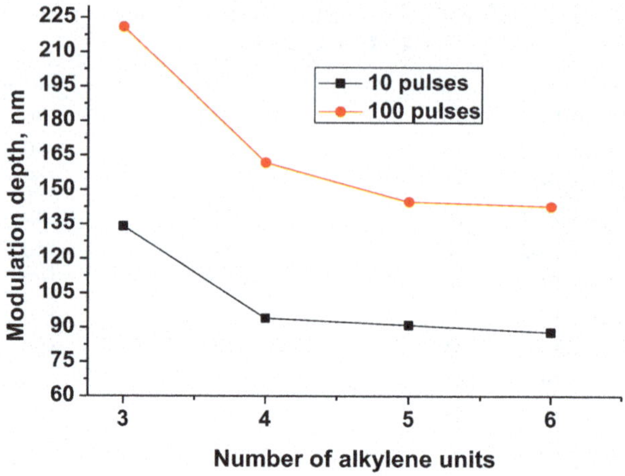

Fig. 5.6 The modulation depth vs. number of alkylene units from side chain

Table 5.2 The modulation depth and texture direction index vs. pulses number and energy fluence for azo-copolyimides **AzoCPI-4–AzoCPI-7**

Polymer	Energy (mJ/cm^2	Pulses number	Modulation depth (nm)	Stdi
AzoCPI-4	45	10	134	0.25
AzoCPI-5	45	10	94	0.33
AzoCPI-6	45	10	91	0.19
AzoCPI-7	45	10	88	0.20
AzoCPI-4	45	100	221	0.29
AzoCPI-5	45	100	162	0.28
AzoCPI-6	45	100	145	0.31
AzoCPI-7	45	100	143	0.34

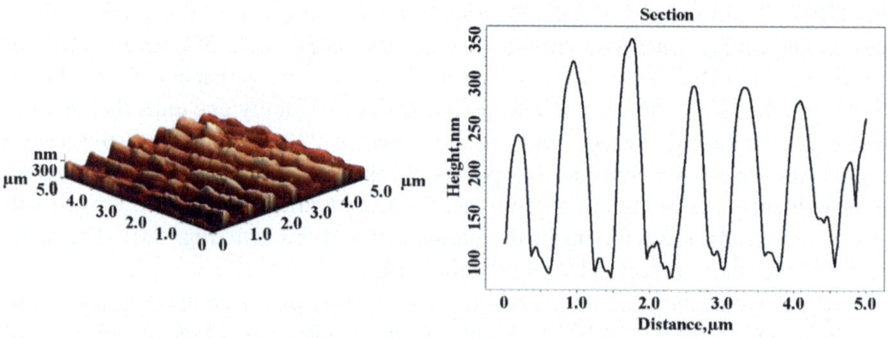

Fig. 5.7 AFM image of the structured surface and corresponding cross-section profile of azo-copolyimide film **AzoCPI-4** after irradiation with 45 mJ/cm^2 by using 100 pulses

5.5 Conclusions

Azopolyimides and azo-copolyimides containing side azobenzene groups have been synthesized by polycondensation reaction of different aromatic dianhydrides (BTDA, 6FDA, or 6HDA) with aromatic diamines containing side azobenzene groups or with a mixture of azodiamines and different aromatic diamines taken in different ratio.

The photochromic studies evidenced that the side azobenzene groups inserted in a polyimide matrix can isomerize and the maximum conversion degree to the *cis*-isomer can reach 69% in solution and 57% in solid state.

The studied polymers showed a good surface structuration capacity in different irradiation conditions. Thus, for an azopolyimide (**AzoPI-6**) with a relatively high content in azobenzene (until 28%), it is necessary a low pulses number (only 5) and a high-density energy (35 mJ/cm^2). A modulation depth of 370 nm with uniform and very well-defined channels was obtained.

Free-standing films were obtained only in the case of azo-copolyimides and in spite of low and a very low content in azobenzene, all azo-copolyimides showed a very good surface relief gratings depending of the nature of used comonomer. In the case of **AzoCPI-1** for a more flexible aromatic diamine was necessary a low energy fluence and a low number of pulses. The more rigid aromatic diamines gave good SRG at high fluence energy irrespective of pulses number and reach a maximum modulation depth of 240 nm.

By increasing the content of azobenzene in polyimide matrix, by using an azodiamine with two-side azobenzene units connected by ether linkages and flexible methylene units the obtained SRGs were regular with uniform channels, irrespective of irradiations conditions. The maximum modulation depth was observed for a small number of methylene units and decreases as well as the number of methylene units increases. The uniform surface relief grating was obtained with a maximum photoinduced pattern depth of 220 nm for polymer **AzoCPI-4** by using an energy fluency of 45 mJ/cm^2 and 100 pulses. A low value of the texture direction index (under 0.35) indicates a preferential orientation of the surface morphology with a good anisotropy.

The surface structuration mechanism is probably based on a process of film surface reorganization and the SRG can be erased by a short thermal treatment at temperatures situated above the T_g value.

Acknowledgements This work was supported by a grant of Ministry of Research and Innovation, CNCS—UEFISCDI, project number PN-III-P4-ID-PCE-2016-0708, within PNCDI III. The authors thank to Dr. Valentin Pohoata, Dr. Ionut Topala (UAIC, Faculty of Physics), and Dr. Ilarion Mihaila (UAIC, CERNESIM) for the access to the laser infrastructure used in this work.

References

1. Wang X (2017) Azo polymers synthesis, functions and applications. Springer, Berlin. https://doi.org/10.1007/978-3-662-53424-3
2. Wang D, Wang X (2013) Amphiphilic azo polymers: molecular engineering, self-assembly and photoresponsive properties. Prog Polym Sci 38:271–301
3. Kumar GS, Neckers DC (1989) Photochemistry of azobenzene-containing polymers. Chem Rev 89:1915–1925
4. Wildes PD, Pacifici JG, Irick G, Whitten DG (1971) Solvent and substituent effects on the thermal isomerization of substituted azobenzenes. A flash spectroscopic study. J Am Chem Soc 93:2004–2008
5. Schanze KS, Mattox TF, Whitten DG (1983) Solvent effects upon the thermal cis-trans isomerization and charge-transfer absorption of 4-(diethylamino)-40-nitroazobenzene. J Org Chem 48:2808–2813
6. Cattaneo P, Persico M (1999) An ab initio study of the photochemistry of azobenzene. Phys Chem Chem Phys 1:4739–4743
7. Shibaev V, Bobrovsky A, Boiko N (2003) Photoactive liquid crystalline polymer systems with light-controllable structure and optical properties. Prog Polym Sci 28:729–836
8. Williams JLR, Daly RC (1977) Photochemical probes in polymers. Prog Polym Sci 5:61–93
9. Nuyken O (1985) Azo polymers. In: Kroschwitz JI (ed) Encyclopedia of polymer science and engineering, vol 2. Wiley, New York, pp 158–175
10. Xie S, Natansohn A, Rochon P (1993) Recent developments in aromatic azo polymers research. Chem Mater 5:403–411
11. Natansohn A, Rochon P (2002) Photoinduced motions in azo-containing polymers. Chem Rev 102:4139–4175
12. Yesodha SK, Pillai CKS, Tsutsumi N (2004) Stable polymeric materials for nonlinear optics: a review based on azobenzene systems. Prog Polym Sci 29:45–74
13. Verbiest T, Burland DM, Jurich MC, Lee VY, Miller RD, Volksen WE (1995) Exceptionally thermally stable polyimides for second-order nonlinear optical applications. Science 268:1604–1606
14. Tsutsumi N, Morishima M, Sakai W (1998) Nonlinear optical (NLO) polymers. 3 NLO polyimide with dipole moments aligned transverse to the imide linkage. Macromolecules 31:7764–7769
15. Sava I, Resmerita AM, Lisa G, Damian V, Hurduc N (2008) Synthesis and photochromic behavior of new polyimides containing azobenzene side groups. Polymer 49:1475–1482
16. Kosela JE, Vapaavuori J, Hautala PA, Faul CF, Kaviola M, Ras RHA (2012) Surface-relief gratings and stable birefringence inscribed using light of broad spectral range in supramolecular polymer-bisazobenzene complexes. J Phys Chem C 116:2363–2370
17. Yadavalli NS, Santer S (2013) In-situ atomic force microscopy study of the mechanism of surface relief grating formation in photosensitive polymer films. J Appl Phys 113:224–304
18. Schab-Balcerzak E, Konieczkowska J, Siwy M, Sobolewska A, Wojtowicz M, Wiacek M (2014) Comparative studies of polyimides with covalently bonded azo-dyes with their supramolecular analoges: thermo-optical and photoinduced properties. Opt Mater 36:892–902
19. Sava I, Burescu A, Stoica I, Musteata V, Cristea M, Mihaila I, Pohoata V, Topala I (2015) Properties of some azo-copolyimide thin films used in formation of photoinduced surface relief gratings. RSC Adv 5:10125–10133
20. Sava I, Stoica I, Mihaila I, Pohoata V, Topala I, Stoian G, Lupu N (2018) Nanoscale analysis of laser-induced surface relief gratings on azocopolyimide films before and after gold coating. Polym Test 72:407–415
21. Ramanujam PS, Holme NCR, Hvilsted S (1996) Atomic force and optical near-field microscopic investigations of polarization holographic gratings in a liquid crystalline azobenzene side-chain polyester. Appl Phys Lett 68:1329

22. Rochon P, Batalla E, Natansohn A (1995) Optically induced surface gratings on azoaromatic polymer-films. Appl Phys Lett 66:136–138
23. Kim DY, Tripathy SK, Li L, Kumar J (1995) Laser-induced holographic surface-relief gratings on nonlinear-optical polymer films. Appl Phys Lett 66:1166–1168
24. Ramanujam PS, Pedersen M, Hvilsted S (1999) Instant holography. Appl Phys Lett 74:3227–3229
25. Leopold A, Wolff J, Baldus O, Huber MR, Bieringer T, Zilker SJ (2000) Thermally induced surface relief gratings in azobenzene polymers. J Chem Phys 113:833–837
26. Baldus O, Leopold A, Hagen R, Bieringer T, Zilker SJ (2001) Surface relief gratings generated by pulsed holography: a simple way to polymer nanostructures without isomerizing side-chains. J Chem Phys 114:1344–1349
27. Rodrıguez FJ, Sanchez C, Villacampa B, Alcala R, Cases R, Millaruelo M, Oriol L (2005) Surface relief gratings induced by a nanosecond pulse in a liquid-crystalline azopolymethacrylate. Appl Phys Lett 87:201914
28. Rodrıguez FJ, Sanchez C, Villacampa B, Alcala R, Cases R, Millaruelo M, Oriol L (2005) Fast and stable recording of birefringence and holographic gratings in an azopolymethacrylate using a single nanosecond light pulse. J Chem Phys 123:204706
29. Kim DY, Li L, Jiang XL, Shivshankar V, Kumar J, Tripathy SK (1995) Polarized laser induced holographic surface relief gratings on polymer films. Macromolecules 28:8835–8839
30. Barrett CJ, Natansohn A, Rochon P (1996) Mechanism of optically inscribed high efficiency diffraction gratings in azo polymer films. J Phys Chem 100:8836–8842
31. Holme NCR, Nikolova L, Ramanuja PS, Hvilsted S (1997) An analysis of the anisotropic and topographic gratings in a side-chain liquid crystalline azobenzene polyester. Appl Phys Lett 70:1518–1520
32. Kim MJ, Seo EM, Vak DJ, Kim DY (2003) Photodynamic properties of azobenzene molecular films with triphenylamines. Chem Mater 15:4021–4027
33. Yager KG, Barrett CJ (2004) Temperature modeling of laser-irradiated azo-polymer thin films. J Chem Phys 120:1089–1096
34. Shirota Y (2005) Photo- and electroactive amorphous molecular materials: molecular design, syntheses, reactions, properties, and applications. J Mater Chem 15:75–93
35. Sava I, Sacarescu L, Stoica I, Apostol I, Damian V, Hurduc N (2009) Photochromic properties of polyimide and polysiloxane azopolymers. Polym Int 58:163–170
36. Ghosh MK, Mittal KL (1996) Polyimides: fundamentals and applications. Marcel Dekker, New York
37. Hergenrother PM (2003) The use, design, synthesis, and properties of high performance/high temperature polymers: an overview. High Perform Polym 15:3–45
38. Liaw DJ, Wang KL, Huang YC, Lee KR, Lai JY, Ha CS (2012) Advanced polyimide materials: syntheses, physical properties and applications. Prog Polym Sci 37:907–974
39. Favvas EP, Katsaros FK, Papageorgiou SK, Sapalidis AA, Mitropoulos AC (2017) A review of the latest development of polyimide based membranes for CO_2 separations. React Funct Polym 120:104–130
40. Yang SY (2018) Advanced polyimide materials: synthesis, characterization and applications, 1st edn. Chemical Industry Press, Elsevier, Amsterdam, Netherlands
41. Ghosh MK, Mittal KL (2018) Polyimides: fundamentals and applications. CRC Press, Boca Raton
42. Sava I, Bruma M, Köpnick T, Schulz B, Sapich B, Wagner J, Stumpe J (2007) New polyimides and polyamides containing azobenzene side groups. High Perform Polym 19:296–310
43. Sava I, Burescu A, Bruma M (2010) Compared properties of polyimides containing pendant azobenzene groups. J Optoelectron Adv Mater 12:309–314
44. Konieczkowska J, Wojtowicz M, Sobolewska A, Noga J, Jarczyk-Jedryka A, Kozanecka-Szmigiel A, Schab-Balcerzak E (2015) Thermal, optical and photoinduced properties of a series of homo and co-polyimides with two kinds of covalently bonded azo-dyes and their supramolecular counterparts. Opt Mater 48:139–149

45. Teboul V, Barille R, Tajalli P, Ahmadi-Kandjani S, Tajalli H, Zielinska S, Ortyl E (2015) Light mediated emergence of surface patterns in azopolymers at low temperatures. Soft Matter 11:6444–6449
46. Schab-Balcerzak E, Skorus B, Siwy M, Janeczek H, Sobolewska A, Konieczkowska J, Wiacek M (2015) Characterization of poly(amic acid)s and resulting polyimides bearing azobenzene moieties including investigations of thermal imidization kinetics and photoinduced anisotropy. Polym Int 64:76–87
47. Konieczkowska J, Janeczek H, Malecki JG, Schab-Balcerzak E (2018) The comprehensive approach towards study of (azo)polymers fragility parameter: effect of architecture, intra- and intermolecular interactions and backbone conformation. Eur Polym J 109:489–498
48. Sava I, Köpnick T (2014) Synthesis and characterization of new diamines containing side substituted azobenzene groups. Rev Roum Chim 59:585–592
49. Sava I, Hurduc N, Sacarescu L, Apostol I, Damian V (2013) Study of the nanostructuration capacity of some azopolymers with rigid or flexible chains. High Perform Polym 25:13–24
50. Sava E, Simionescu B, Hurduc N, Sava I (2016) Considerations on the surface relief grating formation mechanism in case of azo-polymers, using pulse laser irradiation method. Opt Mater 53:174–180
51. Sava I, Burescu A, Bruma M, Lisa G (2011) Synthesis and thermal behavior of polyimides containing pendent substituted azobenzene units. Mater Plast 48:303–307
52. Sava I, Lisa G, Sava E, Hurduc N (2016) Synthesis and characterization of some azo-copolyimides. Rev Roum Chim 61:419–426
53. Sava I (2009) The effect of chemical structure of the surface structuring capacity of some polymers containing azobenzene side groups. Optoelectron Adv Mater 3:718–724
54. Stoica I, Epure L, Sava I, Damian V, Hurduc N (2013) An atomic force microscopy statistical analysis of laser induced azo-polyimide periodic tridimensional *nanogrooves*. Microsc Res Techniq 76:914–923
55. Wang DR, Ye G, Zhu Y, Wang XG (2009) Photoinduced mass-migration behavior of two amphiphilic side-chain azo diblock copolymers with different length flexible spacers. Macromolecules 42:2651–2657

Chapter 6
Structural Modifications of Polymers by Pulsed Electrical Discharges in Liquids

Camelia Miron, Ion Sava, Liviu Sacarescu, Takahiro Ishizaki, Juergen F. Kolb, and Cristian P. Lungu

6.1 Introduction

Polyimides constitute a class of high-performance polymers due to their excellent electrical, thermal, and mechanical properties [1–3]. The ability of a polymer to maintain its integrity yet incorporate new and interesting chemical and physical properties when blended with other materials (such as inorganic nanomaterials) could find many applications. Polyimides (PI) are prepared from inexpensive monomers by several synthetic routes and may serve as an important material for membranes and coatings on different substrates, such as plastics, glass, metals, and silicon [4, 5]. One of the most important applications of polyimide (PI) films is their use as insulators for microelectronic devices, such as for very large-scale integrated and/or ultra-large-scale integrated circuits [6, 7]. The advantages of PI films are good surface planarization characteristics, high electrical breakdown strength, and low dielectric constant, due to the nature of the chemical structure consisting of imide and aromatic rings [8–10]. Insulating materials with dielectric constants as small as possible ($\kappa < 3$, κ: relative permittivity) are required to achieve a maximum

C. Miron (✉)
Center for Low-temperature Plasma Sciences, Nagoya University, Nagoya, Japan
e-mail: camelia@plasma.engg.nagoya-u.ac.jp

T. Ishizaki
Department of Materials Science and Engineering, Shibaura Institute of Technology, Tokyo, Japan

I. Sava · L. Sacarescu
"Petru Poni" Institute of Macromolecular Chemistry, Iasi, Romania

J. F. Kolb
Leibniz-Institute for Plasma Science and Technology, INP Greifswald, Greifswald, Germany

C. P. Lungu
National Institute for Laser, Plasma and Radiation Physics, Bucharest, Romania

C. Miron et al. (eds.), *Carbon-Related Materials*,
https://doi.org/10.1007/978-3-030-44230-9_6

device speed in smaller dimensions [11]. However, the performance of integrated circuit devices is limited by the electrical characteristics of the interconnect (line connecting elements of an integrating circuit). An improvement of the interconnect signal delay, characterized by the resistance-capacitance (RC) time constant, can be achieved by lowering the dielectric constant of the insulator. The result is in an associated reduction of interconnect crosstalk, capacitance, and power dissipation [12]. Signal crosstalk itself is in general determined by the ratio of line-to-line (sidewall) capacitance to total capacitance. The power dissipation in interconnects is proportional to total capacitance, as is the RC delay [12, 13]. As transistor operating voltage continues to scale downward, interconnect crosstalk becomes increasingly important and noise levels must be reduced to avoid spurious transistor turn-on. Since crosstalk is dominated by interconnect sidewall capacitance, process-related solutions such as low-κ materials must be explored. Numerous low-κ organic and inorganic materials spanning a wide range of dielectric constants, from air ($\kappa = 1$) to fluorinated oxides ($\kappa \sim 3.7$) have been explored for use in interconnect systems and replace silicon oxide ($\kappa \sim 3.9$). However, only a limited number have been introduced into production (predominantly with aluminum metallization) due to integration and reliability issues. Integration and reliability issues that have impeded implementation of low dielectrics include thermally or mechanically induced cracking or adhesion loss, poor mechanical strength, moisture absorption, time-dependent behavior, chemical interactions (especially those which may occur during photolithography, etching/cleaning, and dielectric/metal deposition), low electrical breakdown, and poor thermal conductivity. Relative permittivity of less than 2.5 is generally achieved by the introduction of porosity, which further reduces mechanical strength and (for the predominant open pore systems) may increase moisture/chemical adsorption as well. Co-optimization of material, process, and integration architecture will be necessary for successful manufacturing implementation. Insertion of the low-κ dielectric between the metal leads results in a significant reduction in RC delay, extending the performance enhancement curve for at least one technology generation.

Responding to the increasing demand for materials with low-κ, special polyimides (PIs) can be synthesized [11, 14–16]. These PIs can be broadly classified into fully aromatic PIs, semi-aromatic PIs, and fully aliphatic PIs. Aromatic polyimides are usually commercially, such as DuPont's Kapton® and General Electric's Ultem®. Aromatic PIs have very good mechanical and thermal properties which come from strong intermolecular forces between the polymer chains [17]. PIs which contains a charge transfer complex (CTC) consists of two different types of monomers, a donor (i.e., dianhydride) and an acceptor (i.e., diamine) [7–11]. The donor lends electrons from the nitrogen groups to the acceptor with their carbonyl groups, holding them tightly together. Thus, the polymer chains cannot move around very much, and therefore they cannot move around in the whole material. This is the reason why aromatic PIs are so strong and an important challenge is to overcome processing difficulties which usually arise due to infusibility and poor solubility in common organic solvents. They are also stiff and brittle, and it is sometimes necessary to make the polymer softer, so that it may be more easily processed as to lower the dielectric constant. Semi-aromatic PIs may be synthesized from an aromatic monomer, i.e., either a diamine and a dianhydride, but the other part is aliphatic which may minimize the polarizability and decrease the dielectric constant ($\kappa < 3$) [10, 18]. Aliphatic PIs contain aliphatic dianhydrides and diamines and outperform aromatic ones by high processability resulting from higher solubility, high transparency, and low dielectric constant ($\kappa \sim 2.5$) [19, 20]. However, higher flexibility results in lower bond strength and their thermal and dimensional properties are inferior to that of aromatic ones [21]. Hence, PIs derived from aliphatic monomers are only suitable in applications that have less stringent thermal requirements, which may represent an important drawback when used in microelectronics industry.

The color change of PIs from yellow to brown has also limited their application in optical communication and photonic devices [22–29]. The coloration of polyimide was mainly attributed to intermolecular charge transfer (CT) complexes, owing to the molecular aggregation between PI chains in solid state. The fluorescence of PIs is also related to the intermolecular packing and CT nature, where aromatic moieties from diamine act as electron donors and imide moieties act as electron acceptors. Some authors have reported that the CT fluorescence observed for the models mainly originates from the excited intramolecular CT state [23]. However, the insolubility of conventional PIs in organic solvents makes it difficult to distinguish the effects of intra- and intermolecular interaction [22, 25]. Fully aromatic polyimides are known to exhibit fluorescence within the visible region, but the quantum yield of the fluorescence observed is relatively low owing to the strong CT nature of the polyimides and few studies were dedicated to applications of aromatic polyimides for electroluminescent devices [27–29].

Efforts to suppress the CTC interactions and lower the dielectric constant in PIs containing aromatic monomers have included the incorporation of methyl groups on the diamine component leads to an increase in the free volume and a resultant decrease of the molecular packing [18, 30, 31]. The introduction of fluorine-containing groups in the backbone or side chain of the polymer can lead to high hydrophobicity, low polarizability, and oxidation-resistance in comparison with fluorine-free polyimides [32, 33]. Along with the increase in hydrophobicity and increase of the fractional volume, the number of polarizable groups per unit volume is also reduced consequently leading to a decrease of the permittivity [34]. However, the fluoro-PIs possess in general low adhesion strength, low mechanical strength, and low glass transition temperature. The introduction of controlled porosity in materials, typically with pore sizes of 100 Å or less, has been envisioned as a promising method for developing ultra-low dielectric materials. Novel PI hybrid films in which nanoscale polyoxometalate (POM) clusters are covalently linked with the polymer chains or PI clay nanocomposites are synthesized to provide an ultra-low-κ of ~2.5 [35–39]. It was though observed that the introduction of nanopores into PI matrix results in a deterioration of the mechanical and thermal properties of these materials. A series of polyimide/carbon nanocomposite materials using polyimide as the polymer matrix and nanographene sheet and carbon nanotube as fillers, respectively, have improved the dielectric properties, but graphene or graphene oxide with high surface area improved the dielectric properties of polymer most significantly [40–44]. However, the production costs and the preparation time by using solely chemical methods may be an issue.

To overcome some of these problems presented above, plasma discharges have been used for the modification of the properties of different types of polyimides [36–41, 45–50]. Electrical discharges in low pressure reactors and also at atmospheric pressure are often used to modify the surface wettability, roughness, and cross-linking density, but usually a higher surface hydrophilicity is achieved when using these methods [45, 46]. When PI-based materials are used as dielectric layers, permeability also plays an important role because the absorption of small molecules, such as water, modifies the dielectric properties. The absorbed water causes a significant increase in the dielectric constant, even in small concentrations, because of its very large dielectric constant (78.5 at 25 °C) and implicitly may cause considerable degradation of adhesion to different inorganic interfaces in electronic packaging. The reactive species formed in the plasma interact with the polymer to attach functional groups to the surface. After NH3 plasma treatment, the surface properties of polyimide containing multi-wall carbon nanotubes (MWNT) or acid modified MWNTs (M-CNTs) became actually more hydrophilic [46]. The O_2 plasma treatment of the fluorinated polyimide induced a significant decrease of the C–C/C–H bonding, while the N_2 plasma treated fluorinated polyimides determine the defluorination of the polyimide surface and an increase of the surface water absorption [47]. PE-CVD processes allow to deposit fluorocarbon coatings of tuned morphology and properties [48]. The surface treated with the $N_2/He/SF_6/O_2$ plasma increased the surface wettability, while the $N_2/He/SF_6$ plasma increased the contact angle of the sample surface due to the formation of the C–Fx bonding on the surface of the polyimide film [49]. The surface-free energy of the polyimides depends on the plasma parameters used in the experiments. However, these techniques used for the surface functionalization of the polyimide film have some drawbacks due to the excess of energy delivered to the film surface. When the polymer is exposed to the plasma, the energy delivered by the UV radiation or particle bombardment may lead to the random chain scission, molar mass degradation, or crosslinking in polymers [50]. Therefore, the development of novel techniques is necessary for the generation of selective, mono-sort functional groups containing polymer surface, which do not alter the surface or improve the dielectric properties of the materials is necessary for further microelectronics applications.

Trying to avoid processing at low pressure, recently electrical discharges generated at atmospheric pressure in liquids have become one of the most interesting techniques for the functionalization of polymers. The surface of polypropylene films was functionalized with monotype oxygen-based functional groups by underwater plasma (capillary discharge) [51, 52]. It was found that the optical and electrical properties of PI films were also modified after exposure to pulse electrical discharges in distilled water [53–55]. The hydrophobicity of PI films increased and the relative permittivity decreased after treatment by pulsed electrical discharges that are generated in water [53, 54]. Furthermore, plasma in liquids (L-plasma) methods are offering a number of additional capabilities. Liquid plasma-assisted functionalization is faster than solution chemistry and one-step approach [55–64]. The material surface tailoring with different functionalities is also possible, due to a wide range of functional groups that can be generated in liquid plasma by using anionic or cationic solutions. The processes and reactions set in motion are determined by the way the electrical energy that is provided is dissipated in different mechanisms [58–62]. In particular, the development of discharges that are generated with short high voltage pulses (10 ns) is more likely to impart energy primarily on electrons and this way shift the energy distribution function to higher energies accordingly. This will promote processes where higher electron energy is required for collisions, the excitation of radiative states and the development of strong and steep shock fronts. Chemically active species, such as ions, radicals, metastable excited atoms, and UV photons are formed in the discharge which may interact with molecules and bulk-materials, inducing surface modification and functionalization. When using nanosecond voltage pulses, a topographical rearrangement of the polymeric chains is predominantly induced within the bulk structure, while more intense surface modifications are dominating when microsecond pulse discharges are employed to treat the polymer films [53, 65]. A deeper understanding of the liquid-phase plasma is needed for the control of the surface and bulk-materials interactions that can be promoted this way.

6.2 Plasma in Liquids

Pulsed electrical discharges generated at atmospheric pressure in liquids are becoming an increasingly important topic in the field of plasma science and technology. Glow discharge electrolysis and the investigation of the breakdown mechanism of dielectric liquids for high voltage switching were of high interest for plasma in contact with liquids up to about 30 years ago [66, 67]. During the last 15 years, the research on the interactions of plasmas with liquids has broadened to address a variety of application areas, including materials processing. Among the various methods available today for materials processing, techniques for plasma generation in liquids is new, but the approach offers the potential to be highly flexible with regard to their range of applications [56, 68–70]. Different plasmas may be formed in liquids, such as corona, glow discharge, arc or dielectric barrier discharge, on the applied voltage waveform (DC, AC), radio frequency, microwave, or pulsed discharges [71, 72]. The discharges may be initiated at the surface of the liquid, on gas–liquid [73], liquid–liquid [74], or liquid–solid interface [75]. The underlying fundamental processes strongly depend on the specific reactor configuration, operating conditions, and materials used in the discharges. Most of the reports studying plasma in liquids after year 2005 were for nanomaterials synthesis (carbon and metal nanoparticles, nanocomposites) [76, 77].

Several studies were also focusing on the structural modification of polymeric solutions, films, and membranes [50, 53, 78, 79]. Shi et al. have used coplanar barrier discharges as a feasible approach to improve the electrospinnability of polymer solutions and to control the structure of electrospun nanofiber [79]. Pavlinak et al. employed surface dielectric barrier discharges using water-solution electrodes for a permanent surface hydrophilization of polytetrafluoroethylene tubes related to the replacement of a high fraction (more than 80%) of the surface fluorine determined by X-ray photoelectron spectroscopy [70]. The surface of polypropylene films was functionalized with monotype O-functional groups by underwater plasma (capillary discharge) [50, 51]. The achievable content of O-containing functional groups exceeds that of oxygen glow discharge gas plasmas by a factor of two (up to ca. 56 O/100 C). Glow discharge polymerization for synthesis of nitrogen polymers were investigated by Senthilnathan et al. [80]. Electron collision with acetonitrile at the interface initiated the formation of radical monomers, inducing further rearrangement to form low-molecular nitrogen polymers. These radical-rich polymers went through further polymerization, forming stable high-molecular nitrogen polymers. Organic compounds with either unsaturated or high-energy functional groups (e.g., C=C, C=N, and C≡N) form stable free-radical monomers and initiate the polymerization reaction [81]. The chemical properties of polymers produced in plasma differ greatly from those of conventional polymers, as they do not contain regular or repeating units and are enriched with radicalized functional groups [82]. Electrical and optical properties of already synthesized polyimide (PI) films were modified after treatment by pulse electrical discharges in distilled water [54, 55]. It was found that the hydrophobicity of PI films increased and the relative permittivity decreased after treatment by plasmas that are generated in water because surface segregation of CF_3 occurs and/or the concentration of unsaturated bonds at the surface increases [53, 54]. Significant changes in some intrinsic fluorescence features, such as the intensity and position of the emission peak, have been observed during exposure to water plasma. These effects have been considered to correlate with the development of specific chemical interactions between the liquid and the macromolecules, including the formation of hydrogen bridges [55]. In contrast to conventional solution chemistry, plasma in liquids (L-plasma) involves accelerated electrons which contribute to the generation of active chemical species, such as ions, radicals, metastable excited atoms, and UV photons [83]. The discharge may be ignited in different combinations of solvents and solutes, allowing material surface tailoring with different functionalities. The advantage of the plasma in liquids treatment of the polyimides is ensured by the simplicity of the method, not requiring expensive vacuum pumps and being able to be installed at atmospheric pressure.

When water is subjected to a high electric field, highly conductive channels called streamers develop between the electrodes, and the plasma is spontaneously formed in the interelectrode gap [58]. The electrons in these channels efficiently excite and ionize water molecules, producing a high amount of excited species, such as hydrogen, oxygen, nitrogen, and hydroxyl radicals, favoring the introduction of the hydroxyl, carbonyl, and carboxyl groups at the surface, but also peroxide-like functions may be created (Fig. 6.2) [53, 59–63]. Streamer propagation depends on the duration of the applied voltage pulse. For a nanosecond high voltage pulse, it was found that a plasma might be formed directly in the dense liquid phase, without the need for the formation of gaseous bubble by Joule heating first which will then provide a gaseous environment to support a plasma [58, 64].

A velocity up to 2000 km s^{-1} was measured when 10-ns high voltage (32 kV) pulses were applied to a 100-μm diameter electrode [64], while 25 km s^{-1} was the streamer propagation velocity obtained when a 10-μs voltage pulse was applied to generate a discharge [84]. The energy stored in these ionized channels is consumed by radiation, shock waves, and thermal conduction to the surrounding water. The results indicate that also much stronger shock waves can be expected to propagate through the liquid for the discharge generation with shorter high voltage pulses. A previously observed increase of the water conductivity during plasma treatment further suggests the production of chemically active species [85]. The surface of polyimides that are treated by nanosecond pulse discharges in distilled water shows a topographical rearrangement of the polymeric chains occurs. The degree of intermolecular order increases after the treatment in liquid plasma, the distance between the polymer chains is increased or decreased, depending on the plasma parameters and materials used in the discharge [47, 48]. This gives the possibility to generate materials with lower dielectric constant. The porosity of materials may be therefore effectively controlled at the same time with the surface functionalization during the plasma discharge in liquids of different polarity. The results suggest a further study of plasma generated in liquids with respect to associated interactions with organic surfaces and materials.

The aim of this book is to present some recent results obtained for the polyimide films treated by plasma in liquids. Fluorinated polyimides and cobalt-containing fluorinated polyimides were exposed to pulse electrical discharges of different voltage pulse durations (10 ns and 50 μs) and repetition frequencies (10 Hz, 200 Hz, and 7 kHz). Functional changes are explained, in particular of dielectric and optical properties with respect to plasma characteristics and associated chemical processes in the liquid/plasma interface and at the interface with the polymer surface.

6.3 Structural Modification of Plasma Treated Polyimides (Influence of Plasma Parameters on the Structural Properties of the Polyimide Films)

Plasma in liquids may be generated using different reactor configurations and pulse power technology. Electrode materials and arrangement, liquids used in the discharge, and the duration of the high voltage pulses determine different types of plasma, such as spark, arc, or corona discharges [71–75]. As the electron temperature, streamers propagation and velocities depend on these parameters, it is expected that the structural modifications of the polymers treated for different discharge conditions would be different. To verify this hypothesis, aromatic polyimides were exposed to plasma generated in distilled water when using different electrode arrangement and different high voltage pulse durations.

Aromatic polyimides are distinguished from other high performance polymers by the solubility of poly(amidic acid) precursor form, which can be cast into uniform films and quantitatively conversed to polyimide structure. However, fully aromatic polyimides are processed with great difficulties because they are insoluble and infusible, and do not show a glass transition before decomposition. One of the successful approaches to increase solubility and processability of polymers is by introduction of bulky lateral substituents, flexible linkages, nonsymmetric, alicyclic, or nonlinear moieties [85–89]. The introduction of flexible linkages into the backbone of fully aromatic polymers can lead to soluble products; therefore, the synthesis of polyimides containing flexible isopropylidene (6H) and hexafluoroisopropylidene (6F) groups is a promising way to ease processable compounds having high thermal stability.

Aromatic polyimides have been synthesized by solution polycondensation of certain aromatic dianhydrides (hexafluoroisopropylidene diphthalic dianhydride-6FDA) with aromatic diamines (4,4′-diamino-3,3′-dimethyl diphenylmethane-MMDA) by a method previously reported [90]. The polyimide (PI) films were obtained by casting onto glass plate the solution of the PI in DMAc (dimethylacetamide) followed by thermal treatment in a vacuum oven to remove the solvent (in the case of polyimide solution). The thermal treatment started at 60 °C for 4 h, and then the polyimide films were heatedat 100, 150, 200, and 225 °C, consecutively (for 40–60 min at each temperature) (Scheme 1).

Fluorine- and benzene-containing species on a polymer surface are responsible for low surface tension, due to the physical forces which hold molecules together. Highly energetic reactive species that are formed in plasmas generated in water by nanosecond or microsecond high voltage pulses may interact efficiently with the PI films, inducing changes in the surface hydrophobicity. Accordingly, it was studied changes in wettability by measuring the water static contact angle between the liquid meniscus and the polymer surface before and after plasma treatment. Figure 6.1 shows a typical experimental configuration used for the treatment of 6FDA-MMDA polyimide films. Pulsed electrical discharges are generated in a cubic quartz vessel between electrodes immersed in distilled water (5 μS cm^{-1}). Tungsten electrodes arranged in a rod-to-rod configuration were used for the treatment of polymers by spark discharges in water generated by microsecond (50 μs) high voltage pulses [53], while tungsten and stainless-steel electrodes may be arranged in a rod-to-plate configuration for polymer films exposure to corona-like discharges in water generated by nanosecond (10 ns) high voltage pulses [73]. For the both cases, the 40 μm thickness polymer films of 15 × 15 mm^2 were mounted on a PTFE or metal support and immersed in distilled water, facing the interelectrode gap at a distance of 10 mm. The contact angle measured for 6FDA-MMDA polyimide gradually increased with the increase in exposure time, from 74° for the untreated sample to 82°, 87°, and 91° for the 2, 4, and 6 min plasma treated samples when using 50 μs voltage pulses (Table 6.1).

An angle of 74° for the untreated sample increased to 92°, 95°, and 98° after treatment with consecutive 10 ns pulses for 10, 15, and 20 min (Table 6.1).

However, results presented in the following sections show that the mechanism responsible for the increased surface hydrophobicity is different for the treatment with plasma in water generated by different voltage pulse durations. Foremost modifications in surface topography and an increase in the unsaturated compounds at the polymer surface were found to be responsible for the obtained contact angles when

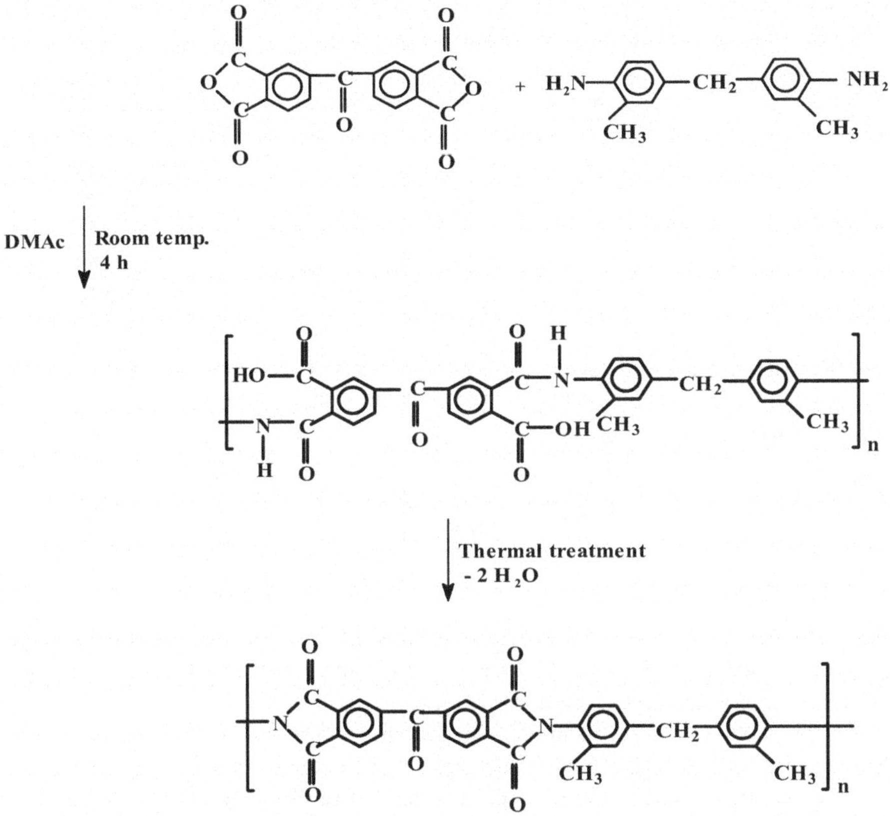

Scheme 6.1 Synthesis of the polyimide 6FDA-MMDA

Fig. 6.1 Schematic of the experimental setup for the generation of pulsed discharges in distilled water for polymer film treatment

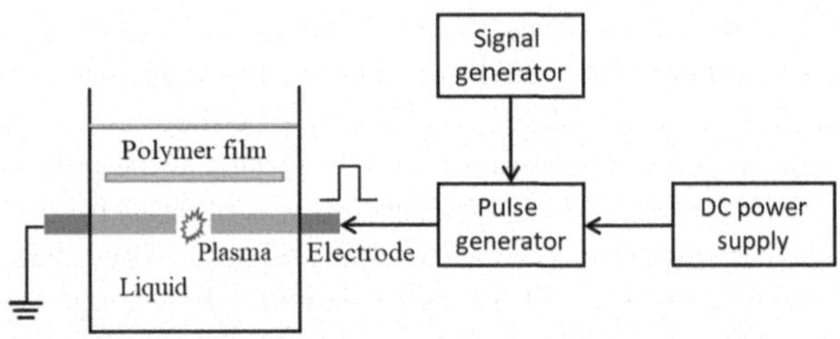

using 50 μs voltage pulses. The fraction of aromatic C–C bonding gradually increased, while the number of aliphatic C–C bonding decreased with the treatment time, resulting in an increase in the surface hydrophobicity of polyimides. The most obvious change in the surface chemical state was given by the fluorine species (Table 6.2). On the surface of the untreated polyimide film could not be detected the fluorine species, but its concentration increased gradually with the discharge time, from 6.0% for the 2 min plasma treated sample to 8.1% in the case of the 6 min

Table 6.1 Contact angle values in (°) of the untreated and PI samples treated with water plasma generated by microsecond and nanosecond high voltage pulses

Pulse voltage duration	Untreated sample [°]	Plasma treatment time [min]	Plasma treated samples [°]
10 ns	74	10	92 ± 3.05
		15	95 ± 3.46
		20	98 ± 3.98
50 μs	74	2	82 ± 2.51
		4	87 ± 2.50
		6	91 ± 2.57

Table 6.2 The element relative atomic concentration of the C, O, N, and F (in atom %) of the PI surface and the rms surface roughness before and after the water plasma treatment

Treatment time [min]	rms [nm]	C [%]	O [%]	N [%]	F [%]
–	2	73.7	21.3	5.0	–
2	14	69.8	17.7	3.9	6.0
4	18	64.8	17.8	3.5	6.6
6	32	64.0	18.2	4.3	8.1

plasma treated PI. This may be attributed to an additional fluorination of the surface induced by the water plasma. Thus, the water plasma yielded surfaces rich in fluorine atoms and with higher roughness [53].

For the plasma treated PI samples the percentage of the C–C bonding gradually decreased to 70.8% (Table 6.3). This shows that the cleavage of a low number of C–C bonds occurs during the plasma treatment of the PI. The relative fraction of C=O was slightly decreased (with ~1%), while the C–O fraction increased after the exposure to the plasma (from ~12.6% to ~17%). The dissociation energy of the C=O bond (11.09 eV) is higher than that of the C–C bond (6.2 eV), explaining the slight modification of the C=O fraction during the discharge. The CF_3 peak at 293.4 eV was identified only in the plasma treated spectra and increased with the discharge time (Table 6.3). This may be the consequence of a segregation process caused by the migration of the CF_3 groups at the polymer surface.

The concentration of CF_3 groups was of 3.0% for the 2 min plasma treated PI, for 4 min treatment increased to 4.1%, and after 6 min, the CF_3 concentration was 4.5%. By angle-resolved XPS was investigated the segregation of the CF_3 group for a better understanding of the surface composition versus the surface energy of the samples. The segregation depths were evaluated starting from inelastic mean free path (imfp) by taking the sampling depth:

$$d = 3\lambda \cos\theta \tag{6.1}$$

where λ is the imfp and θ is the angle between the emitted photoelectrons to the analyzer and the surface normal. The errors are within the range (10–15%). The segregation depths were 4.2 nm, 6 nm, and 8.5 nm for the 2 min, 4 min, and 6 min

Table 6.3 The relative fraction of the C–C, C=O, C–O, and CF_3 peaks of the C1s spectrum before and after the water plasma treatment

Treatment time [min]	C–C [%]	C=O [%]	C–O [%]	CF_3 [%]
–	76.3	9.0	12.6	–
2	71.7	8.3	17.0	3.0
4	72.5	9.0	12.8	4.1
6	70.8	9.1	15.6	4.5

Table 6.4 The relative fraction of the N–C–C and N–C=O peaks of the N1s spectrum before and after the water plasma treatment

Treatment time [min]	N–C–C [%]	N–C=O [%]
–	24.7	75.3
2	–	100
4	–	100
6	–	100

plasma treated samples, respectively. The segregation process is determined by the difference in the surface energy of the polymers film components. The lower surface energy parts will have the tendency to move to the PI surface/air interface.

The nitrogen core-level spectrum of the untreated plasma sample exhibited two components: N–C=O peak at 400.2 eV (75.3%) and N–C–C peak at 398.5 eV (24.7%) (Table 6.4). The concentration of the N–C=O group increased up to 100% with the increasing of the plasma treatment time, while the N–C–C group could not be identified in any spectra of the plasma treated samples. The shift in the nitrogen core-level spectrum at 398.5 eV was explained by some authors as the result of an incomplete imidization of the polymer.

Therefore, it may be assumed that the imidization was incomplete for the untreated plasma film, process which was enhanced under the water plasma action. The degree of imidization increases with the cure temperature (more amide groups cyclodehydrate to imide rings) and attain a maximum value at 240–250 °C. Using a digital thermometer Thermo Tech TT 0986, the water temperature was measured at 20 mm from the core plasma, where the polymer film was placed during the experiments (Fig. 6.10). The water temperature was monitored for 10 min of discharge and it was observed that it did not increase more than 70 °C in all this time. It may be assumed that the imidization of the PI surface may be the consequence of the high-energy excited species which escape from the plasma and reach the film surface. These species might have a much higher temperature than that measured in the surrounding water, inducing the surface modifications mentioned above. The imidization occurred during the plasma discharge determined the formation of PI having the same chemical structure as that of the untreated sample. Therefore, both phenomena, the scission and the imidization of the polyimides, might occur on the film surface under the water plasma treatment.

XPS analysis have shown that the surface chemistry has not significantly changed when polyimides were treated after exposure to nanosecond voltage pulses [65].

The percentage of the aromatic C–C bonding gradually increased to 46% (compared to 41% for the untreated sample) while the aliphatic C–C bonding decreased from 19 to 15% for the treated sample (Fig. 6.2a). This shows that cleavage of aliphatic bonds occurs during plasma treatment. The methyl substitutes of the aromatic diamines might be oxidized after the interaction with reactive species generated in plasma, determining the decrease in the aliphatic bonds and the increase in the oxygen concentration (Fig. 6.2a). Adverse electrostatic repulsion occurs when two π systems stack on top of each other. It can be assumed that the presence of a higher amount of the aromatic C–C bondings at the polymer surface that was observed in the XPS spectra (Fig. 6.2b) determined an increase in the surface hydrophobicity of the polyimide films. However, the concentration of the imidic and C–N bonds remained almost the same after the plasma treatment, suggesting that the main changes occurred to the aromatic diamines structure. This may imply a modification of the surface topography caused by a rearrangement of the nitrogen atom or of the imide ring.

In case of PI films treated by plasma generated with microsecond voltage pulses, the surface chemistry of the polyimide film was also slightly modified, depending on the exposure time of the film to the water plasma.

It may be concluded that plasma in liquids ignited by both, nanosecond and microsecond voltage pulse discharges increase the hydrophobicity of PI films through different reaction mechanisms. Different streamers velocities and type of reactive species formed for the 10 ns and 50 μs discharges may have led to a chemically active species propagation and interaction with molecules and bulk-materials in a different manner, inducing surface modification and functionalization.

6.4 Electrical and Mechanical Properties of Water Plasma Treated Polyimide Films

6.4.1 Polyimide Films Treatment by Nanosecond Pulsed Electrical Discharges in Water

Responding to the increasing demand for synthesis of polyimides with low dielectric constant, polyimide films were treated by plasma in liquids generated by nanosecond and microsecond pulse discharges. The experimental setup described in the previous chapter was also used for the 6FDA-MMDA polyimide films treatment by plasma in water ignited by nanosecond voltage pulses. The frequency and discharge time dependencies of dielectric constant (ε) were investigated for the untreated and plasma treated PIs. The dielectric constant is plotted against frequency in Fig. 6.3, ranging from 10 kHz to 10 MHz. The relative permittivity of the untreated samples was about 3.65 at 1 MHz [54]. The characterization of the frequency dependence of the dielectric constant is a key criterion with regard to microelectronic applications of dielectric materials. The permittivity of PIs decreases, in general, with the increasing of the frequency, due to the frequency dependence of the polarization

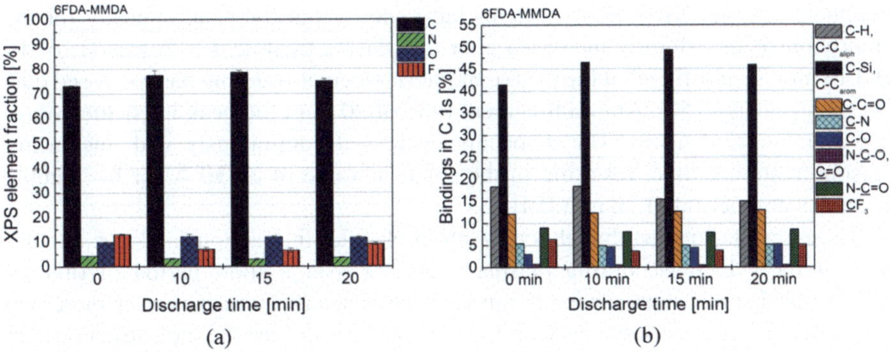

Fig. 6.2 XPS element fraction (**a**) and C1s binding (**b**) of untreated (0 min) and plasma treated 6FDI-MMDA PI films

Fig. 6.3 Dielectric constant vs. frequency for 6FDA-MMDA PI films

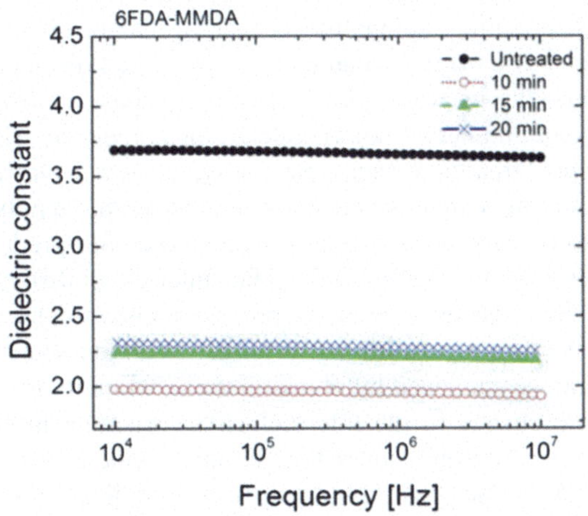

mechanism. The magnitude of dielectric constant is dependent upon the ability of the polarizable units to orient fast enough to keep up with the oscillation of the electric field. The presence of only a few polarizable groups per unit volume usually results in low permittivity of PI films. The variation of the values for the relative permittivity before and after different treatment periods for the PI sample in water plasma are shown in Fig. 6.2. After 10 min of plasma exposure, a remarkably low value of $\varepsilon = 1.92$ at 1 MHz was obtained for the PI film. Longer treatment times of 15 and 20 min resulted in a gradual increase of the relative permittivity to values between 2.2 and 2.27 at 1 MHz (Fig. 6.3). It was previously shown that a decrease in concentration of CF_3 groups also occurs after plasma treatment (Fig. 6.2b).

The lowest concentration of these groups was measured for the 10 min plasma treated sample [65]. The CF_3 peak at 293.4 eV decreased from 6 to 4% after 15 min treatment time, followed by an increase to 5% for 20 min of exposure. This may explain the highest value of the relative permittivity measured for the 20 min plasma treatment sample. XRD measurements have shown that the peak intensity for the amorphous peaks slightly increased after the plasma treatment over periods of 10 and 15 min, suggesting that the degree of intermolecular ordering has increased during the discharge [54]. After a treatment time of 20 min, the peak intensities were, however, reduced again. The d-spacing decreased continuously with increasing plasma treatment time, resulting in the smallest value of 3.350 Å for the sample treated over a period of 20 min [54].

These results suggest that although intermolecular interactions and π–π interactions initially increase during plasma treatment over a short period of time of 15 min, the distance between the chains is also increasing, giving rise to an increased free volume. This is consistent with the observed results for the dielectric constant measurements (Fig. 6.3), suggesting a decrease of permittivity after short plasma treatment time. At longer treatment times (20 min), the distance between chains of the polymer was, however, found to decrease. Also, the reduced peak intensity for both of the broad peak indicates significant change of the intermolecular structure. A possible explanation is that two main effects are being observed. Firstly, the increase in free volume due to increased distances between the chains of the polymer during water plasma treatment causes a drop of relative permittivity since the total number of polarizable atoms is reduced. This effect might be predominant at short treatment times. Secondly, the water plasma treatment may alter the degree and the manner of cross-linking between the polymer chains. This process is likely to occur gradually over time and becomes predominant at longer treatment times, eventually resulting in smaller inter-chain distances and stronger π–π interactions. The occurrence of such a process is consistent with the observed a gradual increase of relative permittivity at longer treatment times. In some cases, π–π interactions can be enhanced by breaking the aromaticity to yield localized pi orbitals. A possible change to the intermolecular structure during plasma treatment may also be reflected in the contact angle measurements previously mentioned, which revealed that surface hydrophobicity was significantly increased during treatment in water plasma [65]. An explanation for the increase of free volume during short treatment times is that higher hydrophobicity reduces the amount of moisture adsorbed, consequently reducing the polarizability of the sample and resulting in lower dielectric constant after exposure to plasma.

6.4.2 *Polyimide Films Treatment by Microsecond Pulse Discharges in Water and Isopropanol*

6.4.2.1 Influence on the Electrical Properties

The experimental configuration used to treat the polyimide film by 50 μs pulse discharges was shown in Fig. 6.4 of the previous chapter. The tungsten electrodes arranged in a rod-to-rod configuration were immersed in distilled water (20 mL) or isopropanol 99.97% (20 mL). The pulse generator (DEI, PVX–4150) was set to deliver pulses of 50 μs at a repetition frequency of 200 Hz, using a delay generator (Stanford Research Systems, Model DG645) [55]. Dielectric relaxation spectroscopy provided information about the storage and dissipation components of the complex permittivity, namely dielectric constant (ε') and dielectric loss (ε'') of the polyimide films in a broad frequency and temperature range. From their variation, thermal transitions correlated with dipolar movements are also evidenced. Dipolar relaxation processes corresponding to thermal transitions are displayed by the steep increase in dielectric constant waves and dielectric loss peaks, which shift to higher temperature with increasing frequency [91–93].

As reported for other polyimides and generally for polymers, dielectric constant, ε', decreases with increasing frequency [94–96]. The dielectric constant being dependent on the ability of polarizable units to align to the alternating electric field, it decreases with increasing frequency, since the orientation of dipole moments needs longer time than that corresponding to applied frequency. It is very important to see that after plasma discharge the dielectric constant of the polyimide films

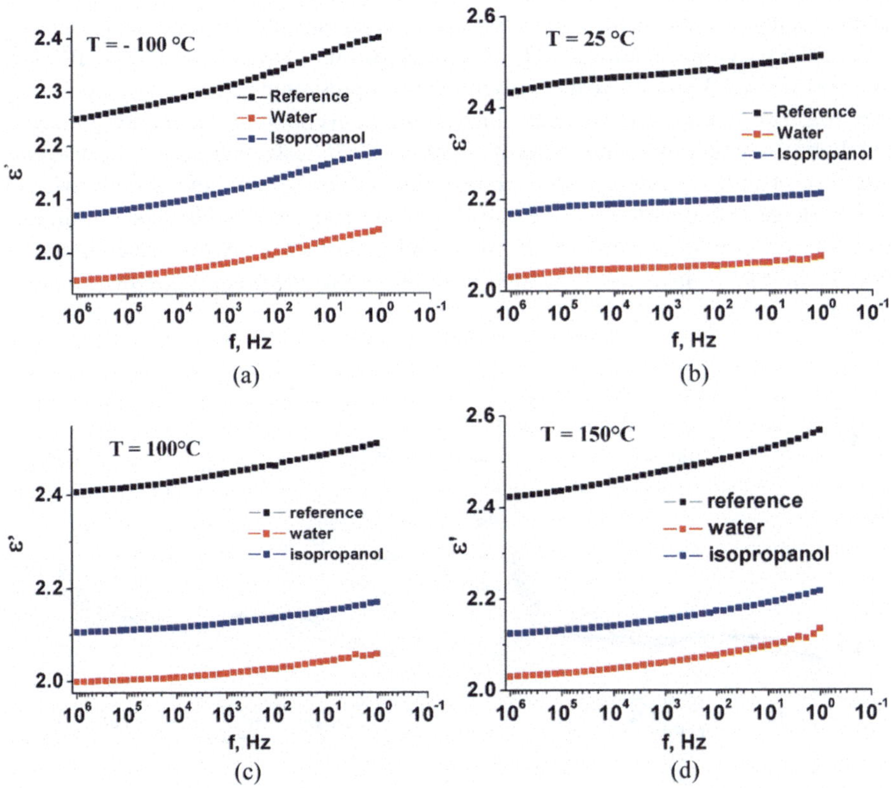

Fig. 6.4 Dielectric constant (ε') versus frequency (f) of the reference and plasma treated polymers in water and isopropanol, at (**a**) −100 °C, (**b**) +25 °C, (**c**) +100 °C, and (**d**) +150 °C

decreases much more in case of water plasma treated samples compared to plasma treated samples in isopropanol.

Figure 6.4 shows that at low (−100 °C) and high (100–150 °C) temperatures, the relative permittivity slightly decreases with the increase in frequency, while at room temperature (25 °C) remains almost constant. In large temperature domains (−150 to 25 °C), the relative permittivity also decreases with the increase in frequency since the thermal energy is not sufficient to activate the dipolar groups (Fig. 6.5) [93, 94]. The relative permittivity showed slightly decreasing values in the range of 25–75 °C, followed by an increase at higher temperatures. At temperatures higher than 200 °C, considerable amount of polarizable units can follow the external field and consequently, the relative permittivity increases sharply with temperature (Fig. 6.5).

The values of the dielectric constant of these polyimide films measured at 25 °C in the frequency domain of 1 Hz–100 kHz are in the range of 2.45–2.51 for reference sample and 2.076–2.18 for the plasma treated polyimide films in both, water and isopropanol (Table 6.5). Plasma treated PI films in water presented lower dielectric constant values compared to those treated in isopropanol, being in the range of 2.076–2.043 and 2.21–2.18, respectively, slightly decreasing with increasing frequency. Table 6.5 shows the values of the dielectric constant (ε') and dielectric loss (ε'') of the polyimide samples at 25 °C and at different frequencies (1 Hz, 10 kHz, and 100 kHz). All plasma treated polyimide films presented a decrease in the dielectric loss. These low values indicate a minimum conversion of the energy to heat in the dielectric material. In the range of −150 to +200 °C, the values of dielectric loss are lower than 0.1, showing that such materials might be exploited as dielectrics in a wide range of temperatures, especially high temperatures [96–98]. The low dielectric constant, in combination with a low dielectric factor, suggest that plasma in liquids method is an efficient method to improve the electrical properties of polymer films.

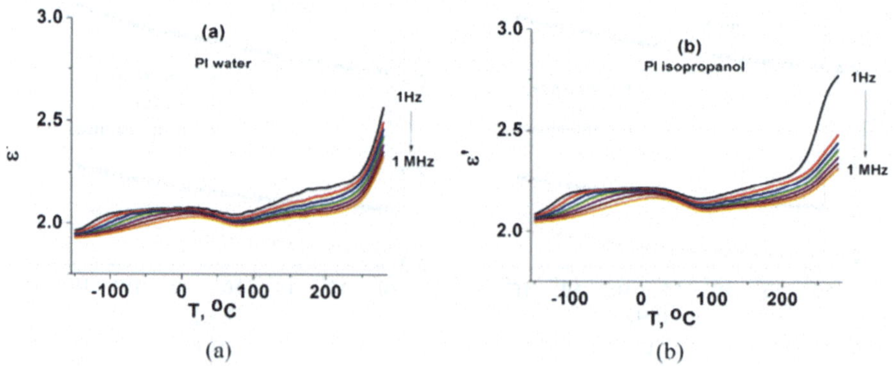

Fig. 6.5 Temperature dependence of dielectric constant for plasma treated PI in (**a**) water and (**b**) isopropanol

Table 6.5 Dielectric parameters (ε', ε'') of polyimide samples at different frequencies

	ε' at 25 °C			ε'' at 25 °C		
Polyimide	1 Hz	1 kHz	100 kHz	1 Hz	1 kHz	100 kHz
Reference	2.51	2.47	2.455	0.0122	0.00627	0.0108
Water	2.076	2.051	2.043	0.00312	0.00282	0.0058
Isopropanol	2.21	2.19	2.18	0.0076	0.0030	0.0074

Reference—pristine polyimide film, water—water plasma treated polyimide, isopropanol—polyimide film treated by plasma formed in isopropanol

Table 6.6 Young's modulus, hardness, and stress–strain values obtained after plasma treatment in water and isopropanol

Polyimide	Young's modulus [GPa]	Hardness [GPa]	Stress [MPa]	Strain [%]
Reference	4.69	0.353	40.92	3.63
Water	5.10	0.442	41.75	3.74
Isopropanol	5.07	0.440	46.17	5.21

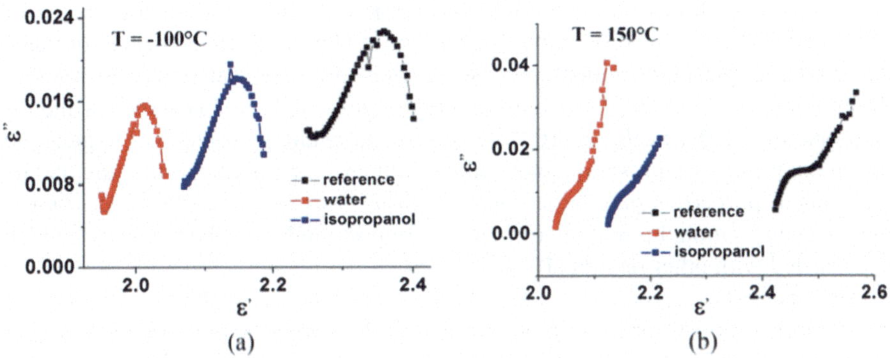

Fig. 6.6 Cole–Cole plots of polyimide films at temperatures of (**a**) −100 °C and (**b**) +150 °C

6.4.2.2 Influence on Mechanical Properties

The influence of the plasma in water or isopropanol on the mechanical properties of the polyimide films was evaluated by measuring the hardness and Young's modulus via nanoindentation (Table 6.6) [99]. Both, hardness and Young's modulus showed higher values than that of the reference sample, suggesting a softer surface and a stiffer interior after treatment. This may be a consequence of molecular rearrangement under the influence of shock waves and streamers formed in plasma. The large change on the arc radius registered in the Cole–Cole plots (Fig. 6.6), pointed to a higher mobility of the chains of plasma treated polyimide film. At high temperature, the shape of the Cole–Cole diagram is linear. This phenomenon accounts for an easier movement of the particle induced by temperature increase, resulting in the enhancement of the dielectric constant. The water plasma treated PI films showed

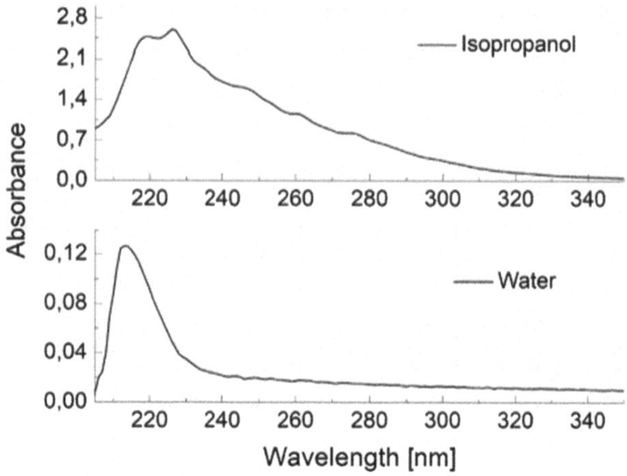

Fig. 6.7 UV–Vis measurement of liquid collected in discharges in water and isopropanol for polyimide film treatment

slightly higher values than that of the PI films treated by plasma in isopropanol. It was previously shown that a topographic rearrangement of polymeric chains of 6FDA-MMDA PI treated by water plasma [65]. The imidization of polymer also occurred after plasma treatment, process which is often associated to an increase in the cure temperature. The dissociation energy of water is 118.82 kcal/mol, while for isopropanol is 105.7 kcal/mol [100]. Therefore, more energy is required to generate plasma in water than in isopropanol and the influence on polymers structure and its mechanical properties may be stronger. With the increase in curing temperature, higher Young's modulus and micro-hardness of polyimides can be measured, which is consisted with other reports [101].

The chain rearrangement can also be the consequence of the development of carbon-rich products from isopropanol and carbon electrodes exfoliation during plasma discharge, such as graphene oxide (Fig. 6.7). The UV–Vis absorption spectrum of plasma formed in isopropanol shows the plasmon peaks near 220 nm and 226 nm, which are attributed to the π–π* transition of aromatic C–C ring [102, 103]. The chain rearrangement can also be the consequence of the development of carbon-rich products from isopropanol and carbon electrodes exfoliation during plasma discharge, such as graphene oxide (Fig. 6.7). The UV–Vis absorption spectrum of plasma formed in isopropanol shows the plasmon peaks near 220 and 226 nm, which are attributed to the π-π* transition of aromatic C–C ring [102]. After reduction, a red shift to 273 nm is observed (Fig. 6.7). This is attributed to n-π^* transition of C–O bonds embedded by exfoliation and intercalation on grapheme [104]. In case of plasma generated in between carbon electrodes immersed in water, only one peak and 214 nm was observed. In this case, graphitic products could have been solely formed during the plasma exfoliation of the carbon electrodes.

The polyimide films were subjected to tensile tests, and the results are shown in Fig. 6.8 and also Table 6.6. The tensile strength and the elongation at break were in the range of 40.9–46.1 MPa and 3.63–5.21%, respectively. It can be observed that the reference and water plasma treated samples shown almost similar values of

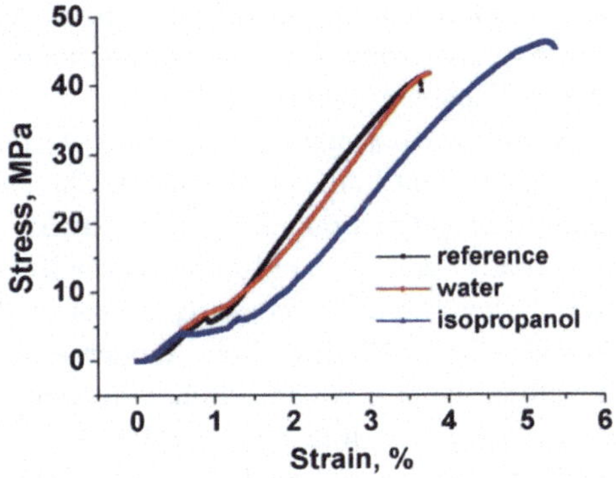

Fig. 6.8 The stress–strain behavior of the polyimide samples

elongation (~3.63%) and strength (~ 40.9 MPa). PI treated in isopropanol has an improvement in elongation (5.21%) and strength (46.1 MPa). This may be explained by the addition of graphene oxide (GO) formed in plasma discharge in isopropanol, in which case an effective load transfer from GO to PI occur (Fig. 6.8). It is unclear why the addition of these nanoparticles leads to this increase in elongation accompanied by an improvement in the strength to failure. Investigations into the cause of these changes in tensile properties are currently ongoing.

6.5 Optical Properties of Water Plasma Treated Fluorinated Polyimides Films Containing Cobalt(II) Chloride Groups

Although PIs have been widely used to microelectronics and aerospace industry, their application in optical communication and photonic devices is limited [22–29]. The coloration of polyimide from transparent yellow to brown was mainly attributed to the intermolecular charge transfer (CT) complexes, owing to the molecular aggregation between PI chains in solid state. The fluorescence of PIs is also related to the intermolecular packing and the CT nature, where aromatic moieties from diamine act as electron donors and the imide moieties act as electron acceptors [22]. Some authors have reported that the CT fluorescence observed for the models is mainly originated from the excited intramolecular CT state [22]. The insolubility of the conventional PIs in organic solvents makes, however, difficult to distinguish the influences of intra- and intermolecular interactions [24]. Fully aromatic polyimides are known to exhibit fluorescence within the visible region, but the quantum yield of the fluorescence observed is relatively low due to their strong CT nature and few studies were dedicated to applications of aromatic polyimides for electroluminescent devices. On the other hand, fluorinated PIs have attracted much attention due to the decrease of the CT interactions by the CF_3-substituted diamines [22, 53]. The introduction of hexafluoropropanedianhydride (6FDA) to the polyimide main chain derived from dianhydrides weakens the intermolecular CT because the bulky substituent group $-C(CF_3)_2-$ prevents the molecular packing of the polyimide film [53].

In case of cobalt modified PIs, it should be mentioned that the cobalt(II) chloride groups may also diffuse preferentially through the amorphous regions, forming charge transfer complexes [22].

Modified fluorinated polyimide materials have different chromic responses to a particular event in the environment. The chromic materials were obtained by introduction of various quantities of cobalt(II) chloride ($CoCl_2$) in the solutions of the polyamidic acid that are further thermally treated to the corresponding imide structure. Scheme 6.2 shows the synthesis method used for the preparation of 6FDA-MMDA-$CoCl_2$ PI film. Low temperature, solution polycondensation reaction of hexafluoroisopropylidene diphthalic dianhydride (6FDA) and 4,4′-diamino-3,3′-dimethyl diphenylmethane (MMDA) was used for preparation of a poly(amidic acid) solution in dimethylacetamide (DMAc). The polyimide films were obtained by casting the poly(amidic acid) solution (which incorporated 15% cobalt(II) chloride hexahydrate) in DMA, onto glass plates and drying at 60 °C over 4 h to evaporate the solvent [105]. The subsequent heating of the precursor films at 100, 150, 200, and 250 °C, consecutively (for 1 h at each temperature) resulted in a final polyimide film. During the thermal treatment, the poly(amidic acid) change the color from blue to turquoise and finally was obtained the Co^{2+}-containing polyimide (PI-Co^{2+}) as a green solid film (Scheme 6.2). The thickness of the PI film used in this study was 100 μm.

Taking into consideration the morality of the chemical formulas which are appointed to the fact that it should be more or less necessary to act in a manner that it should prevent the failure of working with students.

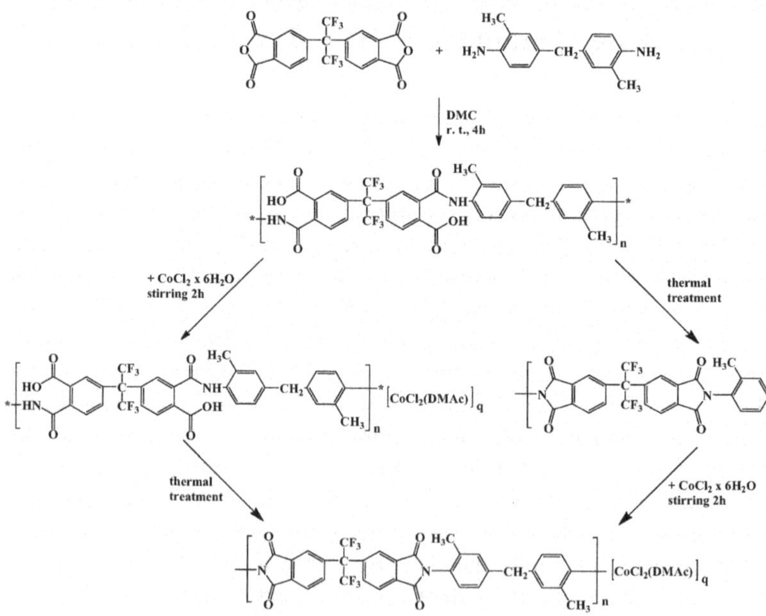

Scheme 6.2 Synthetic paths to Co^{2+}-filled fluorinated polyimides

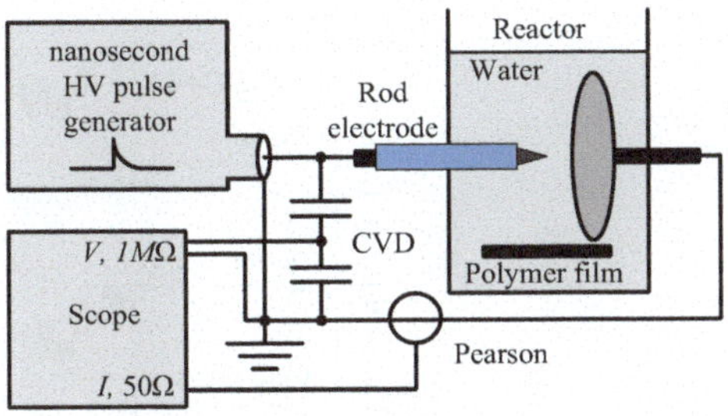

Fig. 6.9 Schematic of the experimental setup for the generation and optical diagnostic of pulsed discharges in distilled water and the exposure of polymer films; CVD—capacitive voltage divider 1:190, Pearson—Pearson Current Monitor Pearson Current Monitor, model: 6585 (500 A_{pik}, 250 MHz)

The Co modified PI films were subjected to plasma treatment using the experimental setup shown in Fig. 6.9. Nanosecond pulsed electrical discharges were generated between two electrodes immersed in 10 mL distilled water (5 μS cm^{-1}). A tungsten rod electrode of 1-mm in diameter and a stainless-steel plate of 15 × 15 mm^2 were arranged in a pin-to-plate configuration. The gap between the electrodes was set to 4 mm [106]. A commercial nanosecond generator (FPG 150-1NK, FID Technology, Burbach, Germany) was used to apply high voltage pulses. The pulse generator can deliver positive high voltage pulses of a rectangular shape with a duration of 10 ns (FWHM).

6.5.1 Fourier Transform Infrared Spectroscopy (FTIR)

The FTIR spectra of polyimide films containing cobalt chloride not-treated and after plasma treatment proved that the characteristic absorption bands of imide's cycle are present in all samples. Figure 6.10 displays representative FTIR spectra of PI/$CoCl_2$ hybrid films (10% $CoCl_2$) in comparison with that of non-treated and after plasma treatment for different period of time (10, 15, and 20 min). Thus, the characteristic imide's ring absorptions appeared in the FTIR spectra of 1720–1725 cm^{-1} (symmetrical C=O imide stretching) and around 1375 cm^{-1} (C–N–C axial stretching), demonstrating the preserving of polyimide cycle [105]. The characteristic absorption bands of hexafluoroisopropylidene are present at 1260 and 1210 cm^{-1}. By increasing the plasma treatment time, the intensity of the band characteristic for the imide cycle remained unmodified, except for the 15 min treated sample. The intensity of the absorption bands at around 1723(C=O) and 1375(C–N) for the 15 min treated sample is lower compared to that obtained for the untreated and plasma treated films for different discharge times. It might be determined by an opening of the imide ring for the 15 min of plasma exposure and a reconstruction when the discharge time is prolonged to 20 min. The intermolecular bonding, such as hydrogen bonding may be renewed, together with some other units of the imide

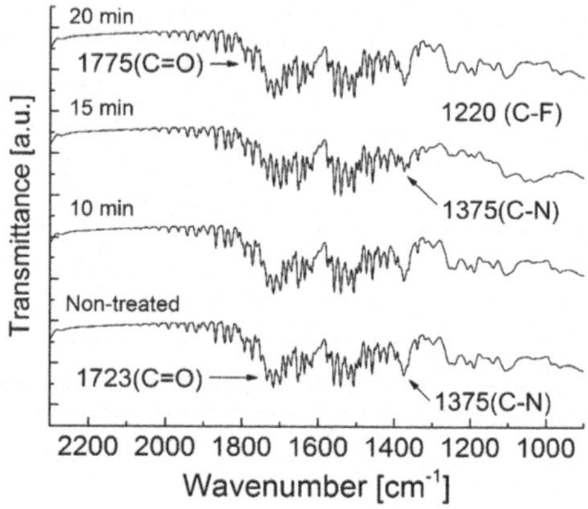

Fig. 6.10 FTIR spectra of non-treated and plasma treatment cobalt chloride-containing polyimide films

cycle (C=O, N–). It may be assumed that highly energetic excited species may reach the polymer surface by means of bubbles and streamers, producing the above-mentioned structural modifications.

Diminishing of the intensity and broadening of the absorption bands were observed in the range of 1100–1400 cm^{-1}. The broadening may be caused by an increase in the strength of the intermolecular interactions, such as C–F bonding or hydrogen bonding belonging to benzene rings. The increase in the temperature may determine the broadening of spectral line and a decrease in the relative intensity. The added thermal energy during the plasma treatment may gradually overcome the intermolecular and intramolecular forces, and less energy is required to excite the vibrational dipoles. XPS results have also shown a gradual increase of the CF_3 peak at 293.4 eV from 1.7 to 4% after 10 min treatment time was observed [106]. This increase may be the consequence of a segregation process caused by the migration of the CF_3 groups at the polymer surface.

However, based on our FTIR and XPS [106] results, we may assume that the active species which are formed under the action of ns-plasma treatment can initiate random hemolytic cleavage of the carbon–nitrogen, carbon–carbon in the imide rings, and carbon–nitrogen bonds in the main chains. These cleavages give rise to chain-free radicals (2,2-hexafluoroisopropylidene diphtalimide, N-phtalimide, 3-methyl-4 amino phenylene, 4-methylene-2-methylaniline) which are recombined and can lead to some characteristic structures (Table 6.7), such as phenyl isocyanate, 2-phenyl-2-phthalimide-hexafluropropane, 2,2-diphenylhexafluropropane, 2,3-dihydro-1,4-phthalazinedione, benzene, and toluene [107].

6.5.2 *UV–Vis Absorption Spectra*

The absorption peaks observed in the transmittance spectra at 538 nm (Fig. 6.11) can be assigned to π–π^* transitions excited from the imide carbonyl (C=O) to the phenyl ring of 6FDA [108]. The cobalt(II) chloride has induced to the charge transfer band

Table 6.7 Possible chemical structures formed after plasma in water treatment of PI films as a consequence of random hemolytic cleavage of the carbon–nitrogen, carbon–carbon in the imide rings, and carbon–nitrogen bonds in the main chains

Compound's chemical name	Chemical structure
Phenyl isocyanate	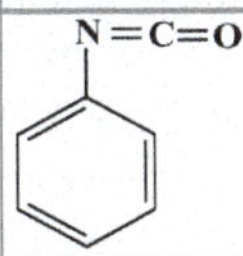
2-Phenyl-2-phthalimide-hexafluropropane	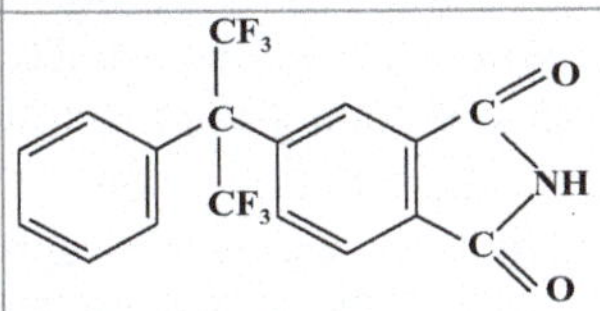
2,3-Dihydro-1,4-phthalazinedione	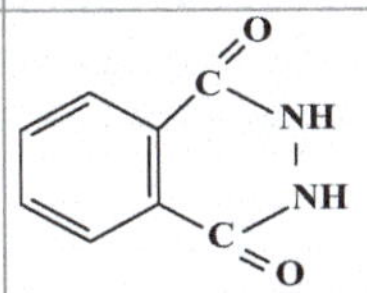
2,2-Diphenylhexafluropropane	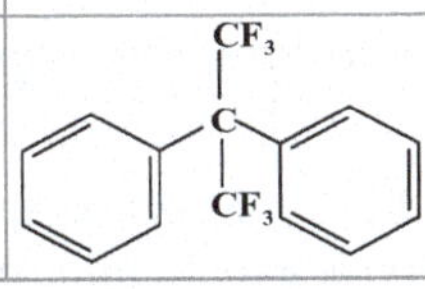

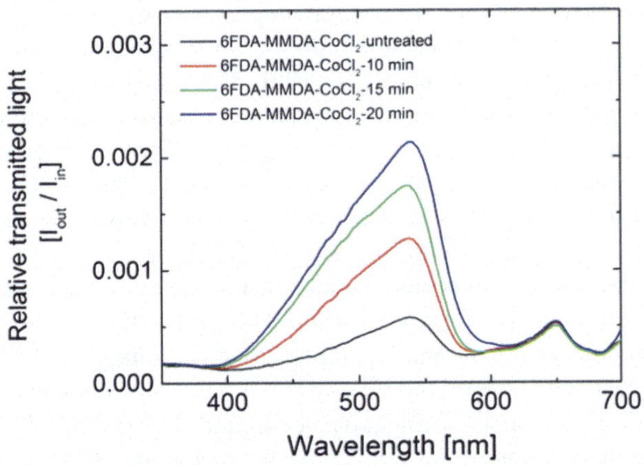

Fig. 6.11 Transmission spectra of 6FDA-MMDA-$CoCl_2$ PI film before and after the plasma treatment

observed at 550–700 nm. The charge transfer band for crystalline cobalt(II) chloride is 670 nm. As the humidity was increased, the peak intensity decreased. This is correlated to the transition change of the weak coordination of the water molecules to the cobalt ion. A decrease in absorbance was observed upon exposure to plasma for all PI samples, indicating an increase in the fluorophore species with an increase of the discharge time (Fig. 6.11). The spectrum obtained was 2 nm blue-shifted for the all plasma treated samples compared to the untreated PI films. This may suggest an ablative process occurring under UV radiation formed in water plasma. The same effect was observed under the photolysis conditions due to polyimide films exposure to a medium pressure mercury lamp in air [109]. A possible explanation for this behavior is the degree of the inter-chain interaction in the form of charge transfer (CT) complexes present in the respective polymers. The strength of these CT complexes depends on the electron affinity of the anhydride moiety and the ionization potential of the diamine moiety. The electron withdrawing ability of the 6F groups serves to increase the electron affinity of the dianhydride and leads to stronger CT complexes. The 6FDA contains a bulky substituent group $-C(CF_3)_2$ between two phenylene groups, reducing the electron density of phenylene groups due to the strong negativity of the fluorine atoms. The steric bulk of these groups may prevent a proper orientation or close enough approach of the chains to one another to form strong CT complex. XPS measurements have shown an increase of the N–C=O, C=O, and CF_3 groups at the polymer surface, indicating a less efficient molecular packing after plasma exposure [106].

6.5.3 Fluorescence Spectroscopy

Figure 6.12a shows the fluorescence spectra of the untreated and water plasma treated PI films for 10, 15, and 20 min. When the film was excited at 450 nm, a red shift of the emission peak at 558 nm was found for the untreated sample, showing marked excitation wavelength dependence of the emission peak. This bathochromic shift is a special phenomenon met in transparent PI films, compared to the case of conventional aromatic PIs [109]. The emission peak was assigned by Hasegawa et al. to the emission of a ground-state intermolecular charge transfer (CT) complex and was considered as the consequence of an increased degree of molecular packing resulted from annealing process [109–111]. They have also stated that the emission from excited ground-state complexes are likely to occur since the energy difference between the ground and excited states of each complex is not monodisperse in amorphous solids, and the red shift of the emission peaks occurred according to the shift in excitation wavelengths. This bathochromic shift may also be due to the formation of the intramolecular hydrogen bonding which change the electronic state of the main chain π-conjugation [112]. The enhancement of intramolecular CT interactions does not significantly occur, and the spectral change mainly originates from enhancement of the van der Waals interaction [113]. The bathochromic shift can be due to the presence of cobalt chloride as was mentioned for other polyimide hybrid which incorporated titania and silica in the macromolecular chain.

The fluorescence peak of the plasma treated samples was shifted to shorter wavelengths compared with the untreated PI film. The emission peak was found at 557 nm for the 10 min treated sample and at 555 nm for the 15 and 20 min of plasma exposed films (Fig. 6.12a). It may be remarked a hypsochromic shift by 1–3 nm and a gradual increase in the emission intensity with the discharge time (Fig. 6.12b). 6FDA-based PIs exposed to pure nitrogen have shown a marked intensity loss of the

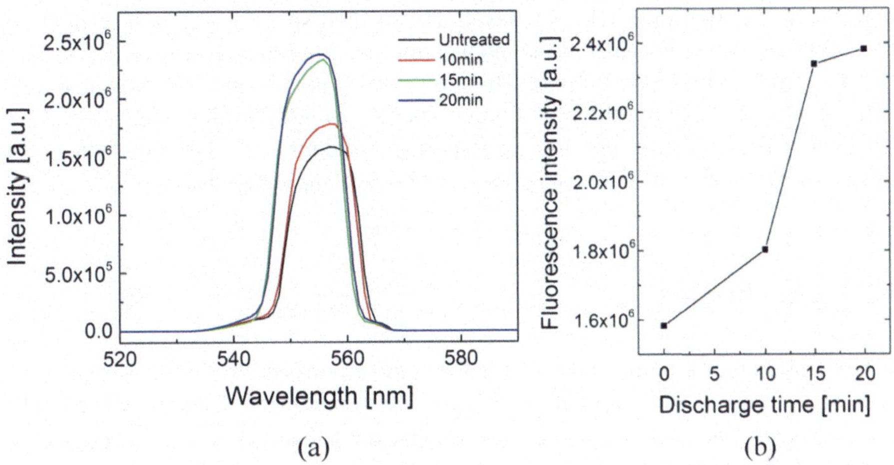

Fig. 6.12 Fluorescence spectra (**a**) and intensity (**b**) of the untreated and plasma treated 6FDA-MMDA-$CoCl_2$ PI films

florescence peaks when nitrogen loaded with ethanol, isopropanol, and water vapor mixtures were used in the experiment [114]. These effects were attributed to the formation of hydrogen bonds between the fluorophore and the alcoholic solvent, between the carbonyl oxygen (CT complex electron acceptor) of the dianhydride moiety of the PI chain and the analyte molecule. The loosened chain packing of the macromolecules caused by the permeation of the analyte molecules through the polymeric network was presented by the authors as a possible mechanism responsible for the emission yield decrease. In our case, water might also get readily adsorbed into the spaces between the chains of the polymer structure, but this cannot explain the increase in the intensity of the fluorescent peak with the discharge time observed after water plasma treatment.

XRD measurements for this PI shown previously in other work have also revealed a decrease in the intermolecular distances and a strengthening of the π–π interactions. Shock waves are generated in plasma [53, 54]. It is well known that variation in pressurization affects the intermolecular and intramolecular CT transitions [112]. It may be assumed that intramolecular rather than intermolecular CT complexes are responsible for the increase in the fluorescence response of the water plasma treated PI samples. The charge transfer in the lowest excited state may also rationalize this blue-shift. Recent studies on polymer blends have shown that the Coulomb interactions between the charge distribution of an excited state chain and that of an adjacent neutral chain can lead to a blue-shift of the emission in the repulsive regime [114, 115]. The collapse of intramolecular hydrogen bonding in blended polymers was found to be responsible for the blue-shifts of the fluorescence peaks [115]. It may be assumed that the intramolecular bonding of the PI films has been destabilized and destroyed under water plasma treatment, inducing changes in the polymer conformational freedom. The slight blue-shift of the emission spectra may be the consequence of the UV radiation formed in plasma [65] which effectively induced modification at the intramolecular level. The excited species may induce random hemolytic cleavage of bonds in the main chain, resulting in new structures, such as 2,3-dihydro-1,4-phthalazinedione. This compound and the new hydrogen bonds created may also explain the blue fluorescence of polyimide films after ns-plasma treatment. These results provide an important guideline for further studies and design of new materials that can be used in high-tech applications.

6.6 Conclusions

Plasma discharge in liquids has emerged as a most powerful tool for materials processing because of its simplicity and being environmentally friendly. Chemically active species, such as ions, radicals, metastable excited atoms, and UV photons, are formed in the discharge which may interact with molecules and bulk materials, inducing surface properties modification and topographical rearrangement of the polymeric chains. Plasma treatment of polyimides may be used to overcome processing difficulties which usually arise due to infusibility and poor solubility in common organic solvents for the purpose of improving the quantum yield of the fluorescence, mechanical strength, or dielectric properties. Plasma treatment on polyimides used in our studies has shown a change in the electrical, optical, and mechanical properties, depending on the nature of liquids used in the discharge and the type of experimental configuration. Thus, polymeric materials treated by the submerged liquid plasma process may find applications in various fields of research, including electronics, nanotechnology, mechanics, biotechnology, and biomedicine. A considerable amount of research is still required to understand fully the plasma reaction in liquids for materials processing.

References

1. Deligöz H, Yalcinyuva T, Özgümüs S, Yildirim S (2006) Electrical properties of conventional polyimide films: effects of chemical structure and water uptake. J Appl Polym Sci 100:810
2. Jeong JY, Babayan SE, Schutze A, Tu VJ, Park J, Henins I, Selwyn GS, Hicks RF (1999) Etching polyimide with a nonequilibrium atmospheric-pressure plasma jet. J Vac Sci Technol A 17:5
3. Isfahani HN, Faghihi K, Hajibeygi M, Bokaei M (2010) New optically active poly(amide-imide)s from *N,N′*-(bicyclo[2,2,2]oct-7-ene-2,3,5,6-tetracarboxylic)bis-L-phenyl alanine and aromatic diamines: synthesis and characterization. Polym Bull 64:633
4. Gonzalo B, Vilas JL, Breczewski T, Perz-jubindo MA, Delafuente MR, Rodriguez M (2009) Synthesis, characterization, and thermal properties of piezoelectric polyimides. J Polym Sci Part A Polym Chem 47:722
5. Zhang QX, Naito K, Tanaka Z, Kagawa Z (2008) Grafting polyimides from nanodiamonds. Macromolecules 41:536
6. Maya EM, Benavente J, de Abajo J (2012) Effect of the carboxylic acid groups on water sorption, thermal stability and dielectric properties of polyimide films. Mater Chem Phys 131:581

7. Sava I, Chisca S (2012) Surface properties of aromatic polymer film during thermal treatment. Mater Chem Phys 134:116
8. Kim Y, Goh WH, Chang T, Ha CS, Ree M (2004) Optical and dielectric anisotropy in polyimide nanocomposite films prepared from soluble poly(amic diethyl ester) precursors. Adv Eng Mater 6:39
9. Ukishima S, Lijima M, Sato M, Takashi Y, Fukada E (1997) A novel cross-linked polyimide film: synthesis and dielectric properties. Thin Solid Films 308:309
10. Lee C, Shu Y, Han H (2002) Dielectric properties of oxydianiline-based polyimide thin films according to the water uptake. J Polym Sci B Polym Phys 40:2190
11. Liang T, Makita Y, Kimura S (2001) Effect of film thickness on the electrical properties of polyimide thin films. Polymer 42:4867
12. Martin SJ, Godschalx JP, Mills ME, Shaffer EO, Townsend PH (2000) Development of a low-dielectric-constant polymer for the fabrication of integrated circuit interconnect. Adv Mater 12:1769
13. Havemann RH, Hutchby JA (2001) High-performance interconnects: an integration overview. Proc IEEE 89:586
14. Chern YT (1998) Low dielectric constant polyimides derived from novel 1, 6-bis [4-(4-aminophenoxy) phenyl] diamantane. Macromolecules 31:5837
15. Chern YT, Shiue H-C (1997) Low dielectric constants of soluble polyimides based on adamantane. Macromolecules 30:4646
16. Meijer HEH, Govaert LE (2005) Mechanical performance of polymer systems: the relation between structure and properties. Prog Polym Sci 30:915
17. Mathews AS, Kim I, Ha CS (2007) Synthesis, characterization, and properties of fully aliphatic polyimides and their derivatives for microelectronics and optoelectronics applications. Macromol Res 15:114
18. Huang JW, Wen YL, Kang CC, Yeh MY (2007) Preparation of polyimide-silica nanocomposites from nanoscale colloidal silica. Polym J 39:654
19. Matsumoto T (2001) Aliphatic polyimides derived from polyalicyclic monomers. High Perform Polym 13:S85
20. Seino H, Sasaki T, Mochizuki A, Ueda M (1999) Synthesis of fully aromatic polyimides. High Perform Polym 11:255
21. Matsuura T, Ishizawa M, Yasuda Y, Nishi S (1992) Polyimides derived from 2,2,-bis(trifluoromethyl)-4,4/-diaminobiphenyl. 2. Synthesis and characterization of polyimides prepared from fluorinated benzenetetracarboxylic dianhydrides. Macromolecules 25:3540
22. Rancourt JD, Taylor LT (1987) Preparation and properties of surface-conductive polyimide films via in situ codeposition of metal salts. Macromolecules 20:790
23. Ukishima S, Iijima M, Sato M, Takahashi Y, Fukada E (1997) Heat resistant polyimide films with low dielectric constant by vapor deposition polymerization. Thin Solid Films 475:308–309
24. Matsuura T, Ando S, Sasaki S, Yamamoto F (1994) Polyimides derived from 2,2/-bis(trifluoromethyl)-4,4/-diaminobiphenyl. 4. Optical properties of fluorinated polyimides for optoelectronic components. Macromolecules 27:6665
25. Halper SR, Villahermosa RM (2009) Cobalt-containing polyimides for moisture sensing and absorption. ACS Appl Mater Interfaces 1:1041
26. Wachsman ED, Frank CW (1988) Effect of cure history on the morphology of polyimide: fluorescence spectroscopy as a method for determining the degree of cure. Polymer 29:1191
27. Takizawa K, Wakita J, Sekiguchi K, Ando S (2012) Macromolecules 45:4764
28. Matsuda S, Urano Y, Park JW, Ha CS, Ando S (2004) Variations in aggregation structures and fluorescence properties of a semialiphatic fluorinated polyimide induced by very high pressure. J Photopolym Sci Technol 17:241
29. Hasegawa M (2001) Semi-aromatic polyimides with low dielectric constant and low CTE. High Perform Polym 13(2):S93

30. Sava I, Chisca S, Bruma M, Lisa G (2010) Study of some aromatic polyimides containing methylene units. Mater Plast 47:481
31. Nica S-L, Hulubei C, Stoica I, Ioanid GE, Ioan S (2013) Surface properties and blood compatibility of some aliphatic/aromatic polyimide blends. Polym Eng Sci 53:263
32. Myung BY, Kim JS, Kim JJ, Yoon TH (2003) Synthesis and characterization of novel polyimides with 2,2-bis[4(4-aminophenoxy) phenyl]phthalein-3′,5′-bis(trifluoromethyl)anilide. J Polym Sci Part A Polym Chem 41:3361
33. Houghham G, Tesoro G, Viehbeck A (1996) Influence of free volume change on the relative permittivity and refractive index in fluoropolyimides. Macromolecules 29:3453
34. Chen H, Xie L, Lu H, Yang Y (2007) Ultra-low-κ polyimide hybrid films via copolymerization of polyimide and polyoxometalates, communication in J. Mater Chem 17:1258
35. Huang JC, Lim PC, Shen L, Pallathadka PK, Zeng KY, He CB (2005) Cubic silsesquioxane–polyimide nanocomposites with improved thermomechanical and dielectric properties. Acta Mater 53:2395
36. Zhang YH, Lu SG, Li YQ, Dang ZM, Xin JH, Fu SY, Li GT, Guo RR, Li LF (2005) Adv Mater 17:1056
37. Zhang Y, Yu L, Su Q, Zheng H, Huang H, Chan HLW (2012) Novel silica tube/polyimide composite films with variable low dielectric constant. J Mater Sci 47:1958
38. Wang HW, Dong RX, Liu CL, Chang HY (2007) Effect of clay on properties of polyimide-clay nanocomposites. J Appl Polym Sci 104:318
39. Wang J-Y, Yang S-Y, Huang Y-L, Tien H-W, Chin W-K, Ma C-CM (2011) Preparation and properties of graphene oxide/polyimide composite films with low dielectric constant and ultrahigh strength via in situ polymerization. J Mater Chem 21:13569
40. Calvert P (1999) A recipe for strength. Nature 399:210
41. Coleman JN, Khan U, Blau WJ, Gun'ko YK (2006) Small but strong: a review of the mechanical properties of carbon nanotube–polymer composites. Carbon 44:1624
42. Yang SY, Ma CCM, Teng CC, Huang YW, Liao SH, Huang YL, Tien HW, Lee TM, Chiou KC (2010) Effect of functionalized carbon nanotubes on the thermal conductivity of epoxy composites. Carbon 48:592
43. Park O-K, Hwang J-Y, Goh M, Lee JH, Ku B-C, You N-H (2013) Mechanically strong and multifunctional polyimide nanocomposites using aminophenyl functionalized graphene nanosheets. Macromolecules 46:3505
44. Yang CY, Chen JS, Hsu SLC (2006) Effects of O_2 and N_2 plasma treatment on 6FDA-BisAAF fluorine-contained polyimide. J Electrochem Soc 153:F120
45. Park SJ, Sohn HJ, Hong SK, Shin GS (2009) Influence of atmospheric fluorine plasma treatment on thermal and dielectric properties of polyimide film. J Colloid Interface Sci 332:246
46. Cheng YY, Ko HH, Chou SC, Ku PS (2014) Study on NH_3 plasma-treated polyimide/MWNT composites on electrical and surface properties. Mater Sci Appl 5:54
47. Lin Y-S, Liu H-M, Tsai C-W (2005) Nitrogen plasma modification on polyimide films for copper metallization on microelectronic flex substrates. J Polym Sci Part B Polym Phys 43:2023
48. Favia P (2012) Plasma deposited coatings for biomedical materials and devices: fluorocarbon and PEO-like coatings. Surf Coat Technol 211:50
49. Oh JS, Gil EL, Kyoung SJ, Lim JT, Yeom GY (2010) Polyimide surface treatment by atmospheric pressure plasma for metal adhesion. J Electrochem Soc 157:D614
50. Friedrich JF, Mix R, Schulze R-D, Meyer-Plath A, Joshi R, Wettmarshausen S (2008) New plasma techniques for polymer surface modification with monotype functional groups. Plasma Process Polym 5:407
51. Joshi R, Schulze RD, Meyer-Plath A, Wagner MH, Friedrich JF (2009) Selective surface modification of polypropylene using underwater plasma technique or underwater capillary discharge. Plasma Process Polym 6:S218

52. Joshi R, Schulze R-D, Meyer-Plath A, Friedrich JF (2008) Selective surface modification of poly (propylene) with OH and COOH groups using liquid-plasma systems. Plasma Process Polym 5:695
53. Miron C, Sava I, Jepu I, Osiceanu P, Lungu CP, Sacarescu L, Harabagiu V (2013) Surface modification of the polyimide films by electrical discharges in water. Plasma Process Polym 10:798
54. Miron C, Zhuang J, Sava I, Kruth A, Weltmann K-D, Kolb JF (2016) Dielectric spectroscopy of polyimide films treated by nanosecond high voltage pulsed driven electrical discharges in water. Plasma Process Polym 13:253–257
55. Sava I, Kruth A, Kolb JF, Miron C (2018) Optical properties of polyimides films treated by nanosecond pulsed electrical discharges in water. Jpn J Appl Phys 57:0102BF
56. Hagoino T, Kondo H, Ishikawa K, Kano H, Sekine M, Hori M (2012) Appl Phys Express 5:035101
57. Senthilnathan J, Liu YF, Rao KS, Yoshimura M (2014) Ultrahigh-speed synthesis of nanographene using alcohol in-liquid plasma. Sci Rep 4:04395
58. Shirafuji T, Noguchi Y, Yamamoto T, Hieda J, Saito N, Takai O, Tsuchimoto A, Nojima K, Okabe Y (2013) Functionalization of multiwalled carbon nanotubes by solution plasma processing in ammonia aqueous solution and preparation of composite material with polyamide. Jpn J Apl Phys 52:125101
59. Kolb JF, Joshi RP, Xiao S, Schoenbach KH (2008) Streamers in water and other dielectric liquids. J Phys D Appl Phys 41:234007
60. Shih K-Y, Locke BR (2011) Optical and electrical diagnostics of the effects of conductivity on liquid phase electrical discharge. IEEE Trans Plasma Sci 39:883
61. Lungu CP, Lungu AM, Chiru P, Pompilian O, Georgescu N, Magureanu M, Mustata I, Velea T, Stanciu D, Predica V (2009) Plasma treatment of levigated materials and saturated waters. Rom J Phys 54:369
62. Marinov I, Starikovskaia S, Rousseau A (2014) Dynamics of plasma evolution in a nanosecond underwater discharge. J Phys D Appl Phys 47:224017
63. Sun B, Sato M, Harano A, Clements JS (1998) Non-uniform pulse discharge-induced radical production in distilled water. J Electrost 43:115
64. Starikovskiy A, Yang Y, Cho YI, Fridman A (2011) Non-equilibrium plasma in liquid water: dynamics of generation and quenching. Plasma Sources Sci Technol 20:024003
65. Miron C, Hulubei C, Sava I, Quade A, Steuer A, Weltmann K-D, Kolb JF (2015) Polyimide film surface modification by nanosecond high voltage pulse driven electrical discharges in water. Plasma Process Polym 12:734–745
66. Hickling A, Ingram MD (1963) Electrochemical processes in glow discharge at the gas-solution interface. Trans Faraday Soc 60:783–793
67. Martin TH, Guenther AH, Kristiansen M (1996) J. C. Martin on pulsed power. Plenum, New York
68. Graham WG, Stalde KR (2011) Plasmas in liquids and some of their applications in nanoscience. J Phys D Appl Phys 44:174037
69. Sunka P (2001) Pulsed electrical discharge in water and their applications. Phys Plasmas 8:2587–2594
70. Pavlinak D, Galmiz O, Zemanek M, Brablec A, Cech J, Cernak M (2014) Permanent hydrophilization of outer and inner surfaces of polytetrafluoroethylene tubes using ambient air plasma generated by surface dielectric barrier discharges. Appl Phys Lett 105:154102
71. Vanraes P, Nikiforov A, Leys C (2016) In: Mieno T (ed) Plasma science and technology: progress in physical states and chemical reactions, pp 457–506
72. Vanraes P, Bogaerts A (2018) Plasma physics of liquids—a focused review. Appl Phys Rev 5:031103
73. Sato M, Tokutake T, Ohshima T, Sugiarto AT (2008) Aqueous phenol decomposition by pulsed discharges on the water surface. IEEE Trans Ind Appl 44(5):1397–1402

74. Khalaf TH, Kadum AH (2010) Analysis of initiation and growth of plasma channels within non-mixed dielectric liquids. Iraqi J Phys 8(11):88–94
75. Li J, Si W, Yao X, Li Y (2009) Partial discharge characteristics over differently aged oil/pressboard interfaces. IEEE Trans Dielectr Electr Insul 16(6):1640–1647
76. Saito G, Akiyama T (2015) Nanomaterial synthesis using plasma generation in liquid., J Nanomater. https://doi.org/10.1155/2015/123696
77. Kareem TA, Kaliani AA (2012) Glow discharge plasma electrolysis for nanoparticles synthesis. Ionics 18:315–327
78. Miron C, Balan M, Pricop L, Harabagiu V, Jepu I, Porosnicu C, Lungu CP (2014) Pulsed electrical discharges in silicone emulsion. Plasma Process Polym 11:214
79. Shi Q, Vitchuli N, Nowak J, Lin Z, Guo B, McCord M, Bourham M, Zhang X (2011) Atmospheric plasma treatment of pre-electrospinning polymer solution: A feasible method to improve electrospinnability. J Polym Sci Part B Polym Phys 49(2):115–122
80. Senthilnathan J, Weng C-C, Der Liao J, Yoshimura M (2013) Sci Rep 3:2414. https://doi.org/10.1038/srep02414
81. Yasuda H, Hirotsu T (1978) Critical evaluation of conditions of plasma polymerization. J Polym Sci Polym Chem Ed 16:743–759
82. Yasuda H (1981) Glow discharge polymerization. J Polym Sci Macromol Rev 16:199–293
83. Banaschik R, Lukes P, Miron C, Banaschik R, Pipa A, Fricke K, Bednarski PJ, Kolb JF (2017) Fenton chemistry promoted by sub-microsecond pulsed corona plasmas for organic micropollutant degradation in water. Electrochim Acta 245:539–548
84. An W, Baumung K, Bluhm H (2007) Underwater streamer propagation analyzed from detailed measurements of pressure release. J Appl Phys 101:053302
85. Liaw DJ, Liaw BY, Hsu PN, Huang CY (2001) Synthesis and characterization of new highly organosoluble poly(ether imide)s bearing a noncoplanar 2,2′-dimethyl-4,4′-biphenyl unit and kink diphenylmethylene linkage. Chem Mater 13:1811–1816
86. Liaw DJ, Chang FC, Leung M, Chou MY, Muellen K (2005) Highly organosoluble and flexible polyimides with color lightness and transparency based on 2,2-bis[4-(2-trifluoromethyl-4-aminophenoxy)-3,5-dimethylphenyl]propane. Macromolecules 38:4024–4029
87. Zhao G, Ishizaka T, Kasai I, Oikawa I, Nakanishi I (2007) Fabrication of unique porous polyimide nanoparticles using a reprecipitation method. Chem Mater 19:1901–1905
88. Ge Z, Fan L, Yang S (2008) Synthesis and characterization of novel fluorinated polyimides derived from 1,1′-bis(4-aminophenyl)-1-(3-trifluoromethylphenyl)-2,2,2-trifluoroethane and aromatic dianhydrides. Eur Polym J 44:1252–1260
89. Zhao G, Ishizaka T, Kasai I, Hasegawa M, Nakanishi I, Oikawa H (2009) Using a polyelectrolyte to fabricate porous polyimide nanoparticles with crater-like pores. Polym Adv Technol 20:43–47
90. Sava I, Chisca S, Bruma M, Lisa G (2010) Comparative study of aromatic polyimides containing methylene units. Polym Bull 65:363
91. Bas C, Tamagna C, Pascal T, Alberola ND (2003) On the dynamic mechanical behavior of polyimides based on aromatic and alicyclic dianhydrides. Polym Eng Sci 43:344–355
92. Kim S-U, Lee C, Sundar S, Jang W, Yang S-J, Han H (2004) Synthesis and characterization of soluble polyimides containing trifluoromethyl groups in their backbone. J Polym Sci B Polym Phys 42:4303–4312
93. Damaceanu MD, Musteata VE, Cristea M, Bruma M (2010) Viscoelastic and dielectric behaviour of thin films made from siloxane-containing poly(oxadiazole-imide)s. Eur Polym J 46:1049–1062
94. Chisca S, Sava I, Musteata V, Bruma M (2011) Dielectric behavior of some aromatic polyimide films. Eur Polym J 47:1186–1197
95. Chisca S, Musteata V, Stoica I, Sava I, Bruma M (2011) Effect of the chemical structure of aromatic-cycloaliphatic copolyimide films on their surface morphology, relaxation behavior and dielectric properties. J Polym Res 20:111

96. Comer AC, Kalika DS, Rowe BW, Freeman BD, Paul DR (2009) Dynamic relaxation characteristics of Matrimid® polyimide. Polymer 50:891–897
97. Damaceanu MD, Rusu RD, Cristea M, Musteata VE, Bruma M, Wolinska-Grabczyk A (2014) Insights into the chain and local mobility of some aromatic polyamides and their influence on the physicochemical properties. Macromol Chem Phys 215:1573–1587
98. Havriliak S, Negami S (1966) A complex plane analysis of α-dispersions in some polymer systems. J Polym Sci Polym Symp 14:99–117
99. Ferreira FAS, Battirola LC, Lewicki JP, Worsley MA, Pereira-da-Silva MA, Amaral T, Lepienski CM, Rodrigues-Filho UP (2016) Influence of thermal treatment time on structural and physical properties of polyimide films at beginning of carbonization. Polym Degrad Stabil 129:399–407
100. Blanksby S, Ellison GB (2003) Bond dissociation energies of organic molecules. Acc Chem Res 36:255–263
101. Lee S-C, Tai F-C, Wei C-H, Yu J-I (2007) ATR-FTIR and nanoindentation measurements of PMDA-ODA polyimide film under different curing temperature. Mater Trans 48:1554–1557
102. Senthilnathan J, Sanjeeva Rao K, Yoshimura M (2014) Submerged liquid plasma–low energy synthesis of nitrogen-doped graphene for electrochemical applications. J Mater Chem A 2:3332–3337
103. Krishnamoorthya K, Veerapandianb M, Kima G-S, Kima SJ (2012) A one step hydrothermal approach for the improved synthesis of graphene nanosheets. Curr Nanosci 8:934–938
104. Lai Q, Zhu S, Luo X, Zou M, Huang S (2012) Ultraviolet-visible spectroscopy of graphene oxides. AIP Adv 2:032146
105. Damaceanu MD, Sava I, Constantin CP (2016) The chromic and electrochemical response of $CoCl_2$-filled polyimide materials for sensing applications. Sensors Actuators B 234:549–561
106. Miron C, Zhuang J, Balcerak M, Holub M, Kruth A, Quade A, Sava I, Weltmann K-D, Kolb JF (2016) Cobalt containing polyimide films treated by nanosecond pulsed electrical discharges in water. IEEE Trans Plasma Sci 44:11
107. Jin X, Huang L, Shi Y, Yang S, Hu A (2002) Thermosetting process and thermal degradation mechanism of high-performance polyimide. J Anal Appl Pyrolysis 64:395
108. Yoshifumi S, Yoshida M, Ando S (2006) J Photopolym Sci Technol 19:297
109. Li Q, Horie K, Yokota R (1997) Absorption, fluorescence and thermal properties of transparent polyimides based on 2, 3, 5-tricarboxycyclopentylacetic acid dianhydride. J Photopolym Sci Technol 10:49
110. Hasegawa M, Kochi M, Mita I, Yokota R (1989) Molecular aggregation and fluorescence spectra of aromatic polyimides. Eur Polym J 25:349
111. Hasegawa M, Shindo Y, Sugimura T, Ohshima S, Horie K, Kochi M, Yokota R, Mita I (1993) Photophysical processes in aromatic polyimides. Studies with model compounds. J Polym Sci Part B 31:1617
112. Wakita J, Ando S (2009) Characterization of electronic transitions in polyimide films based on spectral variations induced by hydrostatic pressures up to 400 MPa. J Phys Chem B 113:8835
113. Carturan S, Quaranta A, Negro E, Tonezzer M, Bonafini M, Maggioni G, Della Mea G (2006) Optical response of 6FDA–DAD fluorinated polyimide to water and alcohols. Sensors Actuators B 118:393
114. Chang MH, Frampton MJ, Anderson HL, Herz LM (2007) Intermolecular interaction effects on the ultrafast depolarization of the optical emission from conjugated polymers. Phys Rev Lett 98:027402
115. Morteani AC, Friend RH, Silva C (2005) Exciton trapping at heterojunctions in polymer blends. J Chem Phys 122:7

Index

C. Miron et al. (eds.), *Carbon-Related Materials*,
https://doi.org/10.1007/978-3-030-44230-9

The manufacturer's authorised representative in the EU is Springer Nature Customer Service Centre GmbH, Europaplatz 3, 69115 Heidelberg, Germany. If you have any concerns regarding our products, please contact ProductSafety@springernature.com

Printed and bound by CPI Group (UK) Ltd, Croydon, CR0 4YY
22/09/2026
02221060-0019